S0-CLX-115

BUILDING AN OPEN SYSTEM

BUILDING AN OPEN SYSTEM

Jacob Slonim
Abraham Schonbach
Michael A. Bauer
Lachlan J. MacRae
Keith A. Thomas

VNR VAN NOSTRAND REINHOLD COMPANY
New York

Library of Congress Catalog Card Number 86-26743
ISBN 0-442-28068-8

Printed in the United States of America

Van Nostrand Reinhold Company Inc.
115 Fifth Avenue
New York, New York 10003

Van Nostrand Reinhold Company Limited
Molly Millars Lane
Wokingham, Berkshire RG11 2PY, England

Van Nostrand Reinhold
480 La Trobe Street
Melbourne, Victoria 3000, Australia

Macmillan of Canada
Division of Canada Publishing Corporation
164 Commander Boulevard
Agincourt, Ontario M1S 3C7, Canada

16 15 14 13 12 11 10 9 8 7 6 5 4 3 2 1

Library of Congress Cataloging-in-Publication Data

Building an Open System

Bibliography; p.
Includes index.
1. Electronic data processing—Distributed processing.
I. Slonim, Jacob, 1945- .
QA76.9.D5064 1987 004'.36 86-26743
ISBN 0-442-28068-8

Preface

A great many books and articles have been written in recent years on open systems, but most of these concentrate on the communications issues, and often tend to leave the impression that once the communications problems are solved, the world of open systems will have arrived. This is by no means the case. One of the aims of this book is to correct that impression by describing the ways in which *all* aspects of the computing environment, in both hardware and software, can promote or impede the realization of the open systems ideal. It is argued, for example, that while new hardware and software are essential to the development of open systems, the new environments must also be able to accommodate the hardward and software, particularly the latter, we use today. This book presents rather more comprehensively than usual the technology of open systems, and attempts to predict the development of that technology in a historical context in which technical, economic, and political problems are intertwined.

JACOB SLONIM
ABRAHAM SCHONBACH
MICHAEL A. BAUER
LACHLAN J. MACRAE
KEITH A. THOMAS

Abbreviations and Acronyms

2PC	Two Phase Commit
2PL	Two Phase Locking
ACID	Atomic, Consistent, Isolated, Durable
AI	Artificial Intelligence
ALPS	Automated Language Processing System
ANSI	American National Standards Institute
ARPANET	Advanced Research Projects Agency Network
BNF	Backus-Naur Form
BSC	Binary Synchronous Communications
CAD/CAM	Computer Aided Design/Computer Aided Manufacturing
CASE	Common Application Service Elements
CATV	Cable TV
CCIS	Common Channel Interoffice Signaling
CCITT	Consultative Committee for International Telephone & Telegraph
CCS	Common Channel Signaling
CICS	Customer Information Control System
CISC	Complex Instruction Set Computer
CMEA	Council for Mutual Economic Assistance
CODASYL	Conference on Data System Languages
COS	Corporation for Open Systems
CPE	Customer Premises Equipment
CPU	Central Processing Unit
CSMA/CD	Carrier Sense Multiple Access with Collision Detection
CSP	Cooperating Sequential Processes
DARPA	Defense Advanced Research Projects Agency
DB2	Database2 (IBM Relational Database Management System)
DBMS	Database Management System
DCE	Data Communication Equipment
DDBMS	Distributed Database Management System
DDL	Data Description Language
DDS	Dataphone Digital Service
DDTS	Distributed Database Testbed System
DEC	Digital Equipment Corp.
DML	Data Manipulation Language
DOD	Department of Defense

DSA	Directory System Agent
DTE	Data Terminal Equipment
DUA	Directory User Agent
ECC	European Economic Community
ECMA	European Computer Manufacturers Association
EIA	Electronic Industries Association
ER	Entity Relationship (Model)
ESPRIT	European Strategic Program for Research
FAX	Facsimile
FCC	Federal Communications Commission
FDM	Frequency Division Multiplexing
GA	Gate Array
Gbps	Gigabits per Second
HP	Hewlett Packard
IBM	International Business Machines
IC	Integrated Circuits
ICOT	Institute for New Generation Computer Technology (Japan)
IDP	Internet Datagram Protocol
IEEE	Institute of Electrical and Electronic Engineers
IMS	Information Management System
IRDS	Information Resource Directory System
ISO/OSI	International Standards Organization/Open System Interconnection
ISDN	Integrated Services Digital Network
KBS	Knowledge Based System
KR	Knowledge representation
Kbps	Kilobits per Second
LAN	Local Area Network
LEONARDO	Low-cost Exploration Offered by the Network Approach to Requirements & Design
LU	Logical Unit
MAILA	Mail Agent
MAP	Manufacturing Automation Protocol
MCC	Microelectronics & Computer Technology Corp.
MF	Mail Forwarder
MIPS	Microprocessors without Interlocked Pipe Stages (Stanford U)
MIPS	Millions of Instructions per Second
MIS	Manager Information System
MISC	Microelectronics & Information Sciences Center (Minneapolis)
MITI	Ministry of International Trade & Industry (Japan)

MNC	Microelectronics of North Carolina
MS	Mail Server
MVS	Multiprogramming with Virtual Storage
Mbps	Megabits per Second
NBS	National Bureau of Standards
NCR	National Cash Register
NIL	Network Implementation Language
NLI	National Language Interfaces
NT1	Network Terminator Type 1
NT2	Network Terminator Type 2
OS	Operating System
OSIE	Open System International Environment Optimization
PABX	Private Automated Branch Exchange
PBS	Private Branch Exchange
PC	Personal Computer
PCI	Protocol Control Information
PCL	Process Control Language
PCM	Pulse Code Modulation
PDL	Program Development Language
PDU	Protocol Data Unit
PLA	Programming Logic Array
PLANET	Programming Language for Networks
PR	Procedural Representation
PSS	Packet Switching Service
PTT	Postal Telephone and Telegraph Authority
RAM	Random Access Memory
RF	Radio Frequency
RISC	Reduced Instruction Set Computer
SDU	Service Data Unit
SEISMU	Software Engineering Institute at Carnegie Mellon University
SFP	Standard Floor Plan
SISAL	Streams & Interaction in a Single Associated Language
SNA	System Network Architecture
SNFS	Sun Network File System
SPC	Software Productivity Consortium (Washington, D.C.)
SQL	Sequential Query Language
SRC	Semiconductor Research Corp.
SWN	System Wide Name
SC	Standard Cell
TAB	Tape Automated Bonding
TDM	Time Division Multiplexing
TC97/SC16	Technical Committee 97, Subcommittee 16

UDS	Universal Directory System
VLSI	Very Large System Integration
WAN	Wide Area Network
XCON	Expert Configurer
XDFS	Xerox Distributed File System for Research in Information Technology

Contents

BUILDING AN OPEN SYSTEM

Chapter 1

Introduction

1.1 DISTRIBUTED SYSTEMS AND OPEN SYSTEMS

In the early 1970s, all but the largest organizations depended on centralized computer centers, through which information flowed in carefully controlled batches, or to which some users were connected with interactive access to very limited processing functions. Networks of computers had begun to appear in some large organizations, but, with the exception of experimental projects such as ARPANET, they reflected the established practices of compartmentalization, hierarchical control, and batch processing. In the mid-1980s, the smallest organizational components—even individuals—can afford interactive computing resources dedicated to their own needs. As well, the operation of most organizations has become more event-driven and dependent on immediate communication and access to information: less oriented to batch processing and more to interactive. This applies both within and between organizations. Decentralized processing has emerged as both affordable and necessary.

Even with this explosion of cheap computing resources, and the changing pattern of organizational behavior, the MIS manager or the systems designer has not escaped the necessity of providing coherent, secure, and reliable systems that serve the needs of the whole organization, not just its individual parts. Activities must be coordinated according to organization-wide rules, and information must be collected, collated, and channeled to appropriate operating units without continual human intervention. In short, information systems that exhibit both the stability of the centralized approach and the flexibility of the decentralized approach must be constructed out of a mosaic of virtually independent processing systems. Two related concepts are involved in a solution to this problem: open systems and distributed systems.

In the broadest context, not restricted to computer systems, an open system is characterized by flows of matter, energy, or information into and out of the system across its natural boundaries. In the context of computer systems, open systems are those characterized by similar flows of data across boundaries between autonomous systems. This entails not only mechanisms to facilitate the flow of data, but also a positive intent on the part of the system's administrators to encourage this sort of traffic.

The term *open system* has been most widely applied to computer systems in connection with the International Standards Organization's Reference Model for Open System Interconnection (ISO/OSI), an internationally accepted template for computer communications standards. An open system in the context of ISO/OSI entails the sharing of control among the members of a system, and direct interaction among such members.

It is possible to create a data-flow between systems via off-line bulk media such as magnetic tape, but control of such transfers is divided into two clearly separated domains: first one system produces a data batch according to its own schedule, then another system processes that batch according to its own schedule. This mode of interaction might at first seem to constitute an open system. Lacking here, however, are the shared control and direct interaction required in transfer by means of two-way continuous channels.

Therefore, in this context, "open system" will refer to a computer system organized and operated in such a way that it can be easily connected to other independently organized and operated computer systems through telecommunication links in order to exchange data on an as-needed basis.

An important aspect of open systems is that specific policies must be instituted and the deliberate actions must be taken in order to enable interconnection and interchange. This cannot happen without the active and positive intervention of a system's administrators. While one can imagine a time when data exchange and intersystem transactions might be almost automatic, perhaps using symbolic directory and translation services embedded in the telecommunications network itself, such capabilities cannot be realized in the present or foreseeable state of the art. Therefore, the implementation of even the simplest degree of open system requires intense cooperation on the part of the administrators of the systems involved.

It should be obvious that the necessary flexibility, especially to cope with growth and technological change, requires that the interfaces between systems be well-defined and widely available, that is, adhere to standards. To a large extent the practicability of open systems is tied to the existence of industry-wide national and international standards; therefore, this book will concentrate heavily on applicable existing and evolving standards.

Another important aspect of open systems is that their components are treated as equals, in what ISO calls peer-to-peer relationships. There are no master-slave relationships, no hierarchy or central control, though there may be distinguished processors within a network that provide unique services. This condition has implications for the way in which coordinated activities can be implemented in open systems.

The open system supports decentralized processing and provides the flexibility to adapt to change, but what supplies the stability for overall control and coordination? There is, at least in a broad sense, a need to take a tightly coupled, centralized information system and "divide it up" over a number of processing sites in such a way that activities can be coordinated and global rules applied, but at the same time that each site be allowed as much local autonomy as possible. This is the concept of a *distributed system*.

Distributed systems rely on the sharing of control and the interactive exchange of information in order to effect overall coordination, much like the communications capabilities required by open systems. In fact, any implementation of a communication protocol that includes the exchange of control or state information as well as data is a distributed system, if only with respect to communications processing. It is possible to conceive distributed systems that are not open systems: systems that are not organized for easy interconnection and data exchange beyond their immediate partners, that utilize non-standard communication techniques, or that are centrally or hierarchically controlled. Such distributed systems support decentralized processing only in a limited way, lacking flexibility and local autonomy. While complete local autonomy—absolute equality among peer systems—is not possible in distributed systems, it can be approached through algorithms built on cooperation and dynamically shared control, exactly as the interfaces within an open system are defined.

The concepts of an open system and a distributed system are separate but strongly related. They are both important in implementing decentralized processing in a contemporary organization. There are three principal problem areas in which both open system and distributed system features are useful:

1. *Data sharing or exchange among separate processing centers.* Open system features such as application protocols are required; however, where parts of the same organization are involved, distributed system features may also be required to enforce stronger rules for coordinating transactions.
2. *Decentralizing from a centralized processing center.* One of the classic reasons for decentralization is to reduce contention, response time, and system overhead costs associated with an overloaded central facility, and to reduce system communication costs, by providing separate processing facilities located closer to their users. Distributed system features are clearly required, but open system features are desirable for growth and flexibility.
3. *Coordinating the activities of separate processing centers.* Where it is necessary to bring some coherence to independent processing

centers (e.g., linking individual workstations into an organization-wide network), open system features are required to effect the flow of data, and distributed system features are required to coordinate activities.

Open systems in support of distributed systems, and distributed systems in the context of open systems, constitute the framework in which practical problems must be approached. Therefore, this book will focus on both and on the relationship between them. Since at present open system features are defined mainly with respect to communication interfaces, the chapters dealing with communication and network services will concentrate on open system issues, whereas the chapters on overall design, OS, and database will concentrate on distributed system issues. Throughout, discussions concerning the trade-offs between local autonomy and overall coordination will examine the interplay between the two concepts.

1.2 THE BOOK: PURPOSE AND STRUCTURE

The past few years have produced a number of books on distributed information systems, devoted for the most part to communications issues and to distributed databases. Such an approach is quite reasonable, since both these areas are critical to the understanding of distributed systems. However, the result of this bias is that many other issues which are of importance to practitioners have been addressed only superficially. We know of no work which provides a ''holistic'' view of distributed systems, still less of open systems. Indeed, there has as yet been little detailed discussion of the notion of an open system and its implications. Nor has there been much direct analysis of the real problems encountered in developing distributed systems in the context of existing systems hardware and software.

What is required, in the authors' opinion, is a global perspective of distributed systems and open systems, encompassing relevant aspects of languages, operating systems, communications, and other tools. The present work hopes to supply such a view.

The argument of the book centers on two key issues:

1. The concept of an open system, and
2. The fact that any practical approach to distributed systems must be built upon and from existing systems.

If there is a single principal theme, it is that practical distributed systems, and thus practical open systems, must be developed in an evolutionary manner from current systems rather than in a revolutionary way: that is to say, we cannot simply discard existing systems.

After this brief introduction we proceed to an overview of those technological advances and trends which form the basis of open systems, present and future. *Chapter 2* reviews trends in a number of areas which have a direct impact on the design and development of open systems, offering brief introductions to many topics which are later treated in detail. There are sections on Hardware, Software, Communications, and Artificial Intelligence.

Hardware. After a general discussion of computer architectures, the section proceeds to specific techniques and devices which are influencing current developments in distributed systems. Among these are the automated layout of integrated circuits, reduced-instruction-set architectures, and microcoding. Apart from the RISC machine, two new types of computer are given special attention: the "multicomputer" and the 32-bit microprocessor.

Software. After a discussion of the general conditions of software in an open, distributed environment, the section focuses on three critical areas. First, distributed database management systems—the heart of any distributed information system. The next topic is Ada, the high-level programming language which is rapidly assuming the status of a de facto standard under the sponsorship of the U.S. Department of Defence. The third issue is the operating system: What qualities are required of an operating system in a distributed environment? The emphasis is on UNIX, which is of course also becoming a de facto industry standard: How adaptable is UNIX to the open environment, and what changes must be made? The section concludes with a survey of commercial and experimental distributed database management systems.

Communications. Rather than attempt a survey of communications tools (for that see Chapter 6), this section concentrates on that single development which holds the greatest promise for the future, the integrated services digital network (ISDN). The ISDN is discussed at greater length later in the book; the overview here is intended to introduce the reader to the basic issues: what it is, what it has to offer, its drawbacks, and its likely course of evolution.

Artificial Intelligence. Since this final section deals with a topic with which many readers are only marginally familiar, the aim is to define the field in general terms, to describe the more important component disciplines of AI, and to show how these disciplines may influence the development of open systems. The section, and the chapter, concludes with a summary of the more important AI research projects.

Chapter 3, on the architecture of open systems, moves from the general to the particular, to the specific techniques whereby distributed and open systems are implemented. The first section discusses the logical architecture, the various ways in which it is convenient to view the structure of an open system. The discussion then concentrates on one of the layers of that structure, the operating system: specifically its role in communications, file management, naming, and database support. The problems of heterogeneous operating systems in a distributed system occupy the next section, together with various prospective solutions: limited-set, portable, and virtual operating systems. The chapter concludes with a more general account of the impact of technological advances on open system architectures: the effect of specific types of new device, and above all the need to devise architectures which can accommodate technological change.

Chapter 4, on directories and dictionaries, is an account of the basic means of locating and identifying all the various elements of an open information system. While such tools are important in any kind of information system, in a distributed system they are crucial, and must be very powerful indeed. To avoid ambiguities, the chapter begins with a set of definitions. The second section describes data dictionaries: their uses and benefits, and the techniques by which they are implemented. Naming techniques as they pertain to data dictionaries are the subject of 4.3, which concludes with descriptions of the naming facilities associated with a number of distributed systems.

Then there is a description of the Universal Directory System developed at Stanford, presented in some detail because it embodies a great number of features desirable in open system directories. Naming systems as they are used in open system directories are discussed in section 4.5, and the following two sections deal with two specific types, machine-oriented and human-oriented naming. Section 4.8 is a brief treatment of the problems of network directories, and 4.10 deals with the management of catalogues in distributed databases. The concluding sections concern the conventions for naming and addressing used in the context of the ISO Reference Model for Open Systems Interconnection (ISO/OSI), and certain implementation issues.

Chapter 5, on the design and implementation of open systems, is essentially a discussion of the principles of software engineering in relation to distributed and open systems. It therefore contains accounts in this special context of each of the phases (beyond requirements) of the software life-cycle: specification, design, implementation, integration and testing, and maintenance. The chapter centers on its most extensive section, which first enumerates what is required of a high-level language in a distributed environment, and then describes and evaluates a number of languages (e.g., Ada, Modula II, Concurrent "C") in the light of those requirements.

Chapter 6 is a survey of the communications tools which are or soon will be available for use in distributed and open systems. The framework of this survey is the ISO/OSI Reference Model. The chapter treats four main subject areas. The first of these is the notion of open systems in communications. An explanation of the open systems concept is followed by a discussion of the ways standards emerge, and then by a description of the ISO/OSI standards themselves.

The second subject is the current state of the art in communications technology. Rather than describing the entire field, we highlight the more significant developments and trends in local area networking, long-haul communications, and network interconnection. The third subject area contains some important practical aspects of communications, including security, performance, and network management. The chapter concludes with an extensive treatment of the integrated services digital network: its benefits, costs, and problems, and what it will offer in both the shorter and longer terms.

Chapter 7, on distributed databases, resembles Chapter 4 in that it redefines terms and techniques well understood in relation to centralized databases for use in distributed systems. A set of precise definitions, presented at the outset, allows quick movement to a discussion of issues pertaining to the DBMS architecture, data models, and conceptual design, all in a distributed environment. Two issues which are of particular importance in a distributed database system, query processing and transaction processing, are given special treatment. Other problems (physical design, administration, definition and description, and communications) are examined in a more cursory manner—most of these are issues which have been treated in other chapters, seen from a DBMS point-of-view. The chapter concludes with a discussion of suitable applications for distributed DBMSs, and a listing of commercial distributed database systems.

Our final chapter, ''Future Developments'' is concerned not so much with the technology of open system as with the way in which open systems will be used in the near future to support various cooperative activities. Our first example is MCC (Microelectronics and Computer Technology Corporation), which is a consortium formed by a number of major American manufacturers of computing equipment. The aim of MCC is to conduct basic research on behalf of all these companies: all will contribute to the input of ideas, and all will share in the output. To support the very ambitious projects of MCC a powerful, distributed, and above all *open* information system is required—it must link and render into a single system all the disparate facilities of the member companies.

The same is true of our second set of examples, national and international research and development projects such as the French ''National

Projects," the Japanese "5th Generation Project," and the "Esprit" projects sponsored by the European Economic Community. These projects are huge, and involve the cooperation not only of companies but also of government or governments. The task of communicating and organizing the vast amounts of information required presents an enormous open system problem in each case.

Our third example, the Corporation for Open Systems, is another type of consortium: in this case an association of cooperating companies is not only faced with the problems of devising an adequate open information system to support their work, the aim of their work is cooperatively to solve just such problems.

Chapter 2

The Technological Basis of Open Systems

2.1 INTRODUCTION

For purposes of discussion the technology relevant to open systems has been divided into five categories: hardware, software, communications, data, and (for lack of a better expression) people. If the last of these seems anomalous, what we mean here is essentially the interaction between people and machines. Since a true open system environment will make available to its users an extraordinary range of tools and services, a *precondition* to the success of such an environment will be a revolutionary change in the technology of user interfaces.

The development of automotive technology provides a useful analogy to this revolution. In early motor cars the driver (user) was responsible for just about everything. Apart from steering and braking, he had to control a wide range of engine functions, such as ignition timing and the richness of the fuel mixture. Since nothing about the machine was very reliable, he had to be able to make impromptu repairs. Every driver had to be a mechanic (a hacker), and all his energies were spent just keeping the thing going, and not very fast at that. Moreover, his automobile was in many details unlike his neighbor's, and to buy a new car was to begin a brand new and quite extended learning curve. But then, only the very hardy would dream of taking such a car any great distance, there weren't very many cars, and they were expensive.

Mass production brought—indeed, it demanded—changes in four areas: cost, reliability, ease of use, and standardization. To move directly from the beginning to the present day, the cars we know are relatively inexpensive, especially in the light of the service they offer. They are admirably reliable—again, relatively speaking. They are easy to drive: just about anybody can drive one with a minimum of training. And they are standardized: learn to drive a Buick and you can drive a Ford, or even a Peugeot.

The low cost of the modern automobile is due directly to mass production, and its reliability to the ''maturity'' of the technology. The main reason the car is easy to drive is that most of its functions have been removed from the responsibility of the driver, who now controls only the steering, acceleration, and braking. Another reason is the standardization of controls.

The driver now has time to do a bit of navigating, and the car can be trusted to take him fair distances without major mechanical incident—the rest can be left to a professional technician. There are lots of cars on the road, and they now travel at high speeds.

In the early days the rules of the road were simple, and they varied from place to place. As the number, range, and speed of automobiles has increased, they have become both more complex and universal: a set of internationally accepted icons now covers most of what we have to do on the road. And speaking of the road, along with the development of the automobile has come an admirably sophisticated highway system, to allow our cars to realize their ability to travel at high speeds over long distances.

The development of the open systems environment must follow a similar course. We already possess a reliable, mature communications technology, and economies of scale, as well as advances in VLSI and packaging techniques, are reducing the cost of processing hardware every month. What we need now is a steady reduction in the number of responsibilities imposed on the users of computer and communications facilities, and a standardization in the interfaces between the user and the many different services (applications) available to him. The open system environment must be "transparent" to its users in the same manner and degree as the mechanical functions of an automobile are transparent to its driver. And finally, the user should be able to move among application services with almost as much ease as a driver can now move from car to car.

In touching the host of specific issues which stem from our two basic areas of concern, we have opened, so to speak, a can of worms. It is impossible *not* to discover real and potential problems where one might never have suspected them. The positive side is that the industry is by no means plunging blindly ahead: researchers are aware of most of these problems. Solutions have already been proposed for many, and others have been addressed at least in principle.

Each of the chapters of this book describes in some detail a different set of key tools in the making of open systems: the present and potential solutions to these problems. It is important to remind oneself from time to time that these are only tools. There is a temptation, in surveying a prospect so grand and elaborate as, say, the proposed integrated services digital network (ISDN) to think of it as an end in itself, rather than a means to an end.

We have attempted to identify those tools which will be most important in the long term, and to give rather more detailed accounts of these on as practical a level as possible. Although our account can't pretend to be exhaustive, it at least attempts to mention every aspect of technology that seems likely to exert an influence on open systems.

Some of the tools have developed apart from the idea of open systems; they are nonetheless essential components, and should be considered in that light. The new 32-bit microprocessors are an example. The open system environment will require inexpensive, substantial processing power at almost every node; 32-bit micros can provide the power of a 1970s mainframe in a personal workstation. Table 2.1, which illustrates the increased capabilities of successive generations of computers, suggests that we can soon expect the power of today's supercomputers in personal workstations. Other tools are answers to specific open systems problems. In general, all aspects of the technology are described in the context of the problems they are to solve.

2.2 HARDWARE

Although computer hardware is not the main subject of this book, recent advances in hardware technology form the real basis of many of the possibilities described herein. One of the main prerequisites of a general open systems environment is the availability of substantial, flexible, and affordable processing power. Over the past few years real breakthroughs in various aspects of hardware design, production, and packaging have caused dramatic reductions in the absolute price of computer power. Moreover, today's personal workstation harbors the power of yesterday's mainframe computer.

On the other hand, the future is built on the past; we cannot leave our history entirely behind us. The open system architectures of tomorrow must incorporate systems which were in place ten years ago, as well as processors and other devices we've not yet imagined. The standards and protocols which bind the open system environment must therefore be bridges between the past and the future. It follows that they themselves must be founded on a clear understanding of the lines of development which connect old and new computer hardware technology.

2.2.1 Computer Architectures

A taxonomic tree (Figure 2.1) of today's computers reveals the growth both of new branches based on parallel processing and of new leaves on all the existing branches. Four areas in particular have advanced rapidly over the last few years [Wallich and Zorpette 1986]:

1. High-speed conventional processors using gate arrays and other custom chips
2. Multiprocessor superminis using off-the-shelf microprocessors

Table 2.1. Five Generations of Computer and Communication Technologies.

GENERATION	FIRST	SECOND	THIRD	FOURTH	FIFTH
Year	1946–56	1957–63	1964–81	1982–89	1990–
Example computers	ENIAC, EDVAC, UNIVAC, IBM 650	NCR-501, IBM 7094, CDC 660	IBM 360, 370; PDP-11, Spectra-70, Honeywell 700, Cray 1, ILLIAC-IV, Cyber-205	Cray XMP, IBM, Amdahl, VAX	Extensive development of distributed computing, merging of telecommunications and compiler technologies, extensive modularity
Computer hardware	Vacuum tubes, magnetic drum, cathode-ray tube	Transistors, magnetic core memories	ICS, Semiconductor memories, magnetic disks, minicomputers, microprocessors	Distributed computing system, VLSI, bubble memories, optical disks, microcomputers	Advanced packaging and interconnection techniques, ultralarge-sale integration, parallel architectures, 3-D integrated-circuit design, gallium arsenide technology, Josephson junction technology, optical components
Computer software	Stored programs, machine code, autocode	High-level languages, Cobol, ALGOL, FORTRAN	Very high level languages, Pascal, operating systems, structured programming, timesharing, DBMS, LISP, computer graphics	ADA, widespread packages, programs, expert systems, object oriented languages	Concurrent languages, functional programming, symbolic processing (natural languages, vision, speech, recognition, planning)
Computer	2 Kilobyte memory	32 Kilobyte memory	8 Megabyte memory	64 Megabyte memory	
Performance	10 Kiloinstruction per second	200 Kips	5 Megainstruction per second	50 Mips	1 Gigainstruction per second to 1 terainstruction per second

SOURCE: Kahn Robert E. "A New Generation in Computing", in Next-Generation Computers (Ed. Edward A. Torrero) IEEE Press, Spectrum Series, 1985, p. 5.

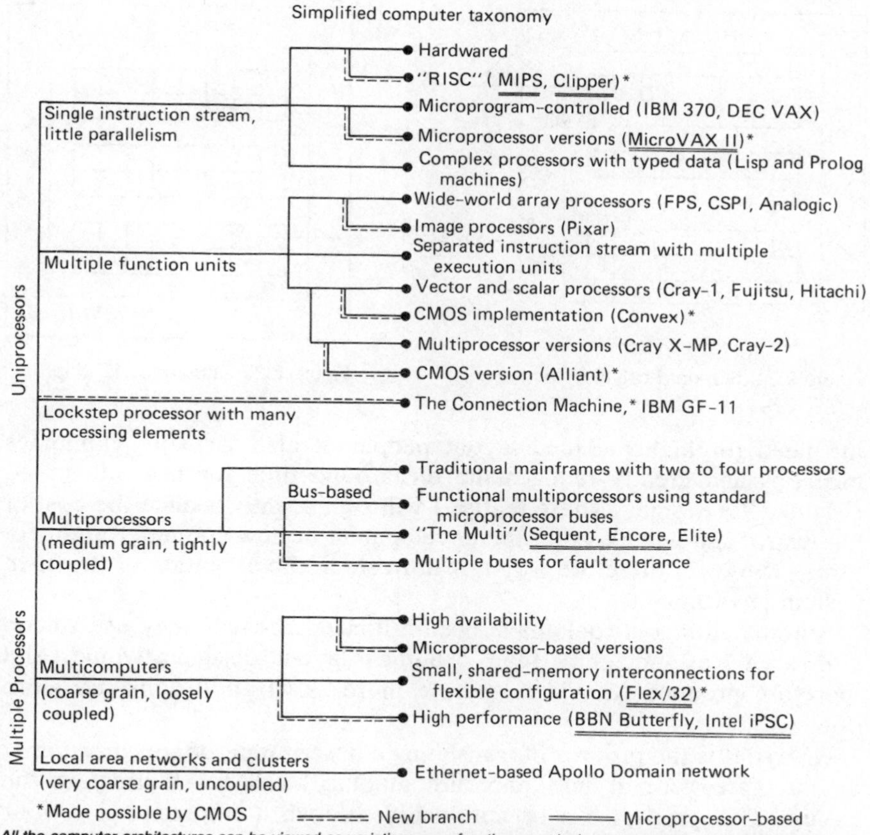

All the computer architectures can be viewed as variations on a few themes: single processor versus multiple processors or single-chip versus multiple-chip implementations, for instance. These variations fan out in a family tree of architectural types, showing where a particular kind of computer may have come from, and where it may be going.

Figure 2.1. Simplified computer taxonomy. (*Source:* Wallich, Paul, and Zonpette, Gienn, "Minis and Mainframes," *IEEE Spectrum*, January 1986, pp. 36—39)

3. Massively parallel computers with all elements operating in synchrony
4. Commercially feasible Reduced Instruction Set Computers (RISC)

2.2.2 Automated Integrated Circuit Layout

Conventional processors are now built by means of automated integrated circuit layout tools, sophisticated programs that translate functional specifications into physical descriptions. By 1990 most integrated circuit layout will probably be automatic. Automating layout not only eliminates

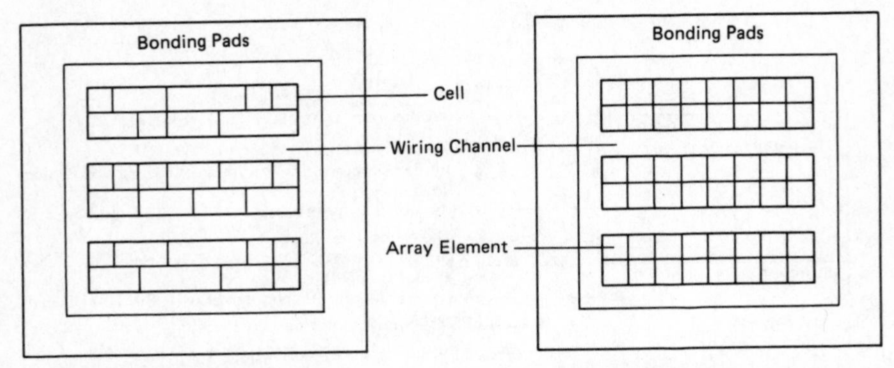

Figure 2.2a. Standard cell IC chip.

Figure 2.2b. Gate array IC chip.

the need for highly skilled layout people, it also virtually eliminates mistakes and greatly reduces the turnaround time for new integrated circuits (IC) designs. These features will significantly reduce the cost of hardware, and will expedite the development of new computer architectures. They will therefore play a major role in the evolution of the open system environment.

Automated layout tools have been criticized because they use silicon chip area less efficiently than a human layout designer would, and therefore produce IC which operate more slowly and consume more power.

IC layout is the process of translating a description, or specification of an integrated circuit into photolithographic mask for fabricating the circuit. This process can be automated through a combination of four basic methods (Figure 2.2):

1. *Standard Cell* (see Figure 2.2a). In this method a large library of predefined cells (small logic elements, such as three input NAND-gates or 4-bit counters) is stored in the layout system. The designer of a circuit tells the system which cells are needed and the kinds of connections that will be required between the cells.
2. *Gate Array* (see Figure 2.2b). A prefabricated chip contains hundreds of thousands of identical cells, such as NAND-gates, arranged in arrays. The designer tells the layout system the logic functions the chip is to perform and the system selects the cells needed and establishes wiring routes among them.
3. *Programming Logic Array (PLA)* (see Figure 2.2c). A chip contains two arrays consisting of NAND-gates or NOR-gates—in series these gates can perform any Boolean logical operation. The designer supplies the PLA layout system with the specifications for general

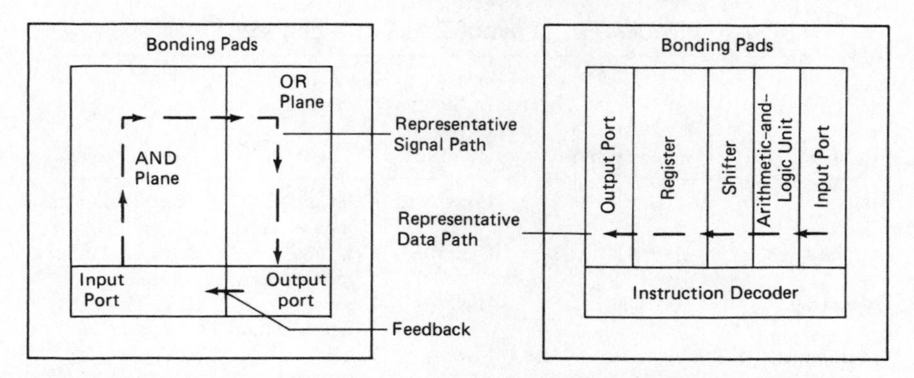

Figure 2.2c. Programmed logic array IC chip.

Figure 2.2d. Standard floor plan.

logical equations and Boolean operations. The system then selects the signals to be included in the array gates to implement the equations.

4. *Standard Floor Plan (SFP)* (see Figure 2.2d). The system contains a generic structure for the chip that is much like a housing developer's basic floor plan for a house, which gives the relative positions of rooms and hallways and the means for connecting utilities to the rooms, but leaves unspecified room sizes and contents. Similarly, the chip floor plan specifies the positions of functional units, bus orientation, and wiring strategy, without predetermining the kinds of elements that will be connected.

Each of these layout tools can accept a functional description of the circuit and may also accept some clues to the placement of logic on the chip. Functional descriptions vary widely among automated layout tools and are tailored to particular tools and the needs of designers (see Table 2.2). A good match between problem and strategy yields an efficient implementation. Standard cell layout, for example, does well on shift registers, PLA layout produces excellent control circuits, and standard floor plan layout yields efficient processor circuits.

2.2.3 32-Bit Microprocessors

Not long after the first 4- and 8-bit microprocessors were introduced in the early 1970s, when 16-bit chips were just visible on the horizon, designers realized that the progression would eventually lead to microprocessors which access and manipulate addresses and data 32 bits at a time. General-purpose minicomputers and mainframes had already more or less settled on that width, as a good compromise between speed, which

Table 2.2. Layout Tools at a Glance

TOOL	APPLICATIONS	PERCENT OF SPACE FOR WIRING	INPUT	MAJOR USERS
Manual	All	Low	Cells and connections	Virtually universal
Standard cell	All	High (up to 80)	Cells and connections	Bell Labs, others
Gate array	All	High	Logic specification	IBM, Motorola, others
Programmed logic array	Combinational logic, state machines	Low	Logic equations	IBM
Standard floor plan	Frequently designed large structures (such as processors)	Low	Functional parameters	Caltech, IBM, Bell Labs (experimental)

Note: Place-and-route systems—that is, standard cell and gate-array layout systems—are versatile but use chip area extravagantly for wiring. Programmed logic arrays and standard floor plans by contrast conserve area almost as much as manual layout but they have limited applications.

SOURCE: Trimberger, Stephen. "Automating Chip Layout," IEEE Spectrum, June 1982.

favours long words for faster access to large blocks of data, and efficiency, which prefers limited widths to avoid wasting bits.

Applications for the 32-bit micros are developing slowly. Workstations, multiprocessing machines, and real time systems seem to be the major users at present, but the consensus among computer people who have used the chips is that the surface has only been scratched. Beside the Motorola chip MC68020 [MacGregor and Robinstein 1986], four other 32-bit microprocessors are commercially available at the beginning of 1986:

- The NS32332 from National Semiconductor in Santa Clara, CA
- The WE32100 from AT&T in Holmdel, NJ
- The 80386 from Intel Corp of Santa Clara, CA [El-Ayat and Agarwal 1985]
- The Z8000 from Zilog Inc. of Campbell CA [Phillips 1985].

Dozens of other 32-bit designs are either awaiting commercial introduction or have been developed only for use in systems sold by the designer. Among the U.S. companies known to have developed or to be developing 32-bit microprocessors are Digital Equipment Corp., Fairchild Camera & Instrument, Advanced Micro Devices, Inc., Hewlett Packard Co., Data General Corp., NCR Inc., INMOS, UNISYS Corp., IBM Corp., McDonnell Douglas Corp., and TRW Inc. In Japan 32-bit development is under way at Hitachi Ltd., NEC Corp., Fujitsu Microelectronics Inc., Toshiba Corp., and Mitsubishi Corp.

The latest generation of 32-bit chips shows the influence of the mainframe and minicomputer architectures. This influence is apparent in several ways, including pipelining, memory management, floating-point arithmetic, and processing rates. Pipelining increases the throughput of the machines by preparing one or more instructions for execution while another instruction is being executed. The MC68020 and WE32100 can operate simultaneously on three 32-bit words, which may represent three different instructions or part of a single instruction. For many of the applications run on the new 32-bit chips, the three-stage pipeline is considered a reasonably optimal depth.

Memory management (virtual memory) techniques are used on the 32-bit chips. These techniques allow programmers to write programs much larger than could fit in main memory space available to the CPU. The programs are simply stored on a secondary device, and portions of the program are swapped into main memory as needed. Two memory-management units, the AT&T WE32101 and the Intel 80386, allow a choice between paging, segmenting, and a combination of both modes. Memory management hardware is also an alternative to cache memory for on-chip implementation, a route taken by both Intel with its 32-bit 80386 and Zilog with its 32-bit Z8000. The advantages of on-the-chip memory management include the ease with which a high degree of parallelism between translations and memory access may be provided, and small propagation times between the memory-management unit and processor.

Caches are relatively small, high speed memories installed between processors and their main memory. The theory behind caches is that a significant proportion of the CPU time spent in running typical programs is tied up in executing loops. Thus the caches are good if an instruction to be executed is the next sequential instruction, or will be at some relatively small number of instructions back. This concept is known as "locality of reference". Therefore a high-speed memory (25-ns static RAM) large enough to contain most loops should increase processing rates.

Floating-point co-processors for the new chips are implemented as extensions of the architecture, just as on mainframe and minicomputer CPUs. With a properly modified compiler, the floating-point unit seems to the user to be an addition to the CPU instruction set. All addressing modes of the CPU can be applied to the floating point registers.

In processing rates the 32-bit chips either approach or surpass some superminis. Although the execution rate of a CPU, commonly measured in millions of instructions per second (MIPS), is a straightforward concept in theory, it turns out to be exceedingly hard to define in practice because of the wide variety of instruction-set and CPU architectures. Nonetheless, by most estimates, for typical number crunching applications the NS32032 comes close to the 1-MIPS performance widely attributed to the VAX 11/780. Processing rates, however, depend heavily on the type of application. For example, a design calling for a great deal of communications with secondary storage or one involving multiple users would need high performance input/output subsystems like high speed buses and disk drives to match the performance of a supermini.

The prices of the 32-bit microprocessors are still high: all are in the $300 to $500 range. By contrast, the price of the MC68000, which has been available for five years, is about $10 to $20 [Zorpette 1985].

2.2.4 Microcode

To the general problem of how to deal with the complexities of the complex instruction sets of specific machines there are two basic approaches: microcode and reduced-instruction-set (RISC) machines. Microcoding has been proposed as a means of mediating between different machine architectures—at points where no common ground exists in higher-level instructions—and thus of interconnecting such architectures.

Most modern computers, including microprocessors, execute their instructions in microcode, which is yet another level of coding below the individual bits of the instructions fetched from memory. These microinstructions give individual and more detailed control over the registers, data paths, and arithmetic and logic units of the processor. Microcode serves to simplify both the programming and design of the central processing unit (CPU), since the programmer (or the compiler program) is not required to consider which operations can or cannot proceed in parallel or exactly what bit to set on. Because they control the CPU at a much greater level of detail, microinstructions are typically much wider than macroinstructions.

Microcoding simplifies hardware design because designers can compose many high-level operations from a small number of microinstructions rather than adding entirely new hardware for each new instruction. They can implement a multiplication instruction, for example, as a series of addition and bit-shifting microinstructions rather than building a hardware multiplier.

Instructions in microcode can often run faster than they would if they were spelled out in machine code or in macroinstructions, because microinstructions have access to internal registers and other machine resources that are not visible at the higher level. Furthermore, because microcode is built into the CPU, the microinstructions can be fetched from the control store much faster than macroinstructions can be retrieved from main memory.

Because microinstructions can be used to build specialized high-level functions from simpler low level ones, they are also often used to hide the low-level architecture of a computer from a programmer.

Microinstructions, however, also present problems. The difficulty of writing them is an obvious one. Wide-word microcode is particularly hard to write and debug, because it involves many actions occurring in parallel with hard-to-follow instructions between them. So-called vertical microcode, which allows much less parallelism, is easier to write but less efficient. Another difficulty is that microcode adds a level of indirection to the execution of instructions. Unless hardware is added to decode instructions and fetch their microcode quickly, a microcode computer will run more slowly than one that executes instructions directly. With the advent of very fast memories this may no longer be a problem, especially with wide-word microcoding, where most instructions require only a single line of microcode: in this case microcoding is simply an alternative method of getting the correct data to the processor's internal control lines and is equivalent to a programmed logic array or any other control structure.

2.2.5 RISC Machines

Reduced-instruction-set computers (RISC) are an alternative to high performance single-processor designs. A RISC eliminates complex instructions in the interest of speed and simplicity. The current RISC movement began in the early 1970s, when studies at IBM Corp. revealed that most of a computer's time was spent loading data from memory into registers and storing them back into memory. Program branches, arithmetic operations, loop instructions, or transfers to different parts of a program occurred relatively frequently. Many other instructions, on the other hand, were used infrequently.

Nonetheless, most modern large computers have enormous instruction sets. The DEC VAX, for example, has 304 instructions, each of which can be used with one or more of the computer's 18 addressing modes to operate on one or more of its 20 data types. An extreme example is provided by Honeywell, which in the early 1970s offered an instruction set for one of its mainframe computers that reduced every COBOL verb to a single corresponding machine instruction. Such an approach gives rise to the following problems:

1. Large instruction sets require complex and potentially time-consuming hardware steps to decode and execute them.
2. Complex machine instructions may not match high-level language instructions exactly, and may therefore be of little use.

3. Rich instruction sets present an overwhelming choice to language compiling programs, which may therefore be incapable of finding the correct specialized instruction to carry out a particular high level function.
4. Since complex machine instructions often have intricate execution sequences and instruction size affects programs, their use can be difficult to optimize.
5. Instruction sets designed with specialized instructions for several high-level languages will carry excess baggage when executing any one language.

Computers based on RISC principles, on the other hand, make choices simpler for compilers—programs that translate high-level-language programs into series of machine-code instructions for a computer to execute. Moreover, since their instructions break operations down into primitive units, it is relatively easy for optimizing programs to combine operations and make the software faster.

Thorough critical examinations of reduced instruction-set computer concepts can be found in [Colwell et al. 1985, Patterson 1985 and Wallich 1985]. Table 2.3 contains a summary of Complex Instruction Set Computer (CISC) and RISC approaches.

The three pioneering RISC machines are the IBM 801 (creatively named for the research group's building number), RISC I (University of California at Berkeley), and MIPS (Microprocessor without Interlocked Pipe Stages, Stanford University). Colwell and his colleagues have derived a set of core features of RISC machines. They propose the following elements as a working definition of a RISC:

1. *Single-cycle operation* facilitates the rapid execution of simple functions that dominate a computer instruction stream and promotes a low interpretive overhead.
2. *Load/store design* follows from a desire for single-cycle operation.
3. *Hardwired control* provides for the fastest possible single cycle operation. Microcode leads to slower control paths and adds to interpretive overhead (see explanation below).
4. *Relatively few instructions* and addressing modes facilitate a fast, simple interpretation by the control engine.
5. *Fixed instruction format* with consistent use eases the hardwired decoding of instructions, which again speeds control paths.
6. *More compile-time effort* offers an opportunity to move static run-time complexity into the compiler.

Another example of a RISC implementation is the very impressive Fairchild Clipper microprocessor chip set, which consists of five chips,

Table 2.3. Summary of CISC and RISC Approaches to Computer Design

	COMPLEX (CISC)	REDUCED (RISC)
Registers	8–16 general register floating point	16–32 general register, optional floating point
Data type	Bytes . . . double precision floating point decimal, byte strings, page tables, queues, etc.	Bytes . . . integers, floating point (optional) decimal (?) byte strings (?) (software processing of OS data)
Instructions	Correspond to data type, instructions assist OS and run-time utilities	Load/store general registers, operations on data types in registers
Instruction formats	Variable-length, many types: load/store, R:=R OP R, R:=Mem OP R, M:=M, M:=Mem OP Mem	Fixed-length, two main types: load/store, R:=R OP R
Encoding	1 instruction = 1 statement	1 instruction = 1 operand or 1 operation
Design objective	Minimum program length, maximum work per instruction	trade-off program length; minimize time required to execute instruction.
Implemented	Microprogrammed processor; slow primary memory and fast clock; instructions taking variable times; complex pipeline; larger implementation may result in longer design time	Hard-wired processor and software; fast processor and fast cache for instructions; instruction take one clocktime; simple pipeline
Caching	Useful	Essential for instructions
Compiler design	Should stress finding correct instructions	Should stress optimal ordering
Philosophy	Move useful software functions into hardware, including diagnostics, hardware changes	Move all functions to software

and executes instructions at roughly the same rate as the VAX 8600. The processor being built by MIPS Computers Inc. exploits work on simple, fast architectures done at Stanford University. (The U.S. Defence Advanced Research Projects Agency has let a contract for gallium arsenide version intended to reach 100 million instructions per second.)

2.2.6 Multicomputers

A new class of high-performance supermini computers is the multicomputer—a shared-memory multiprocessor system with a single bus for interconnection and a cache memory for each processor to reduce delays caused by bus traffic. Computers of this type have been introduced by Elite (6 processors), Encore (20 processors), Sequent (up to 12 processors), and Geac (the Concept 9000, with up to 8 processors). A basic distinction among multiprocessor systems is between those which use bus-based architectures and those which do not. The bus-based group can be divided into two subcategories: tightly coupled and loosely coupled systems. The tightly coupled systems, sometimes simply called multiprocessors, have both a multiprocessor and a common, or global memory. The processors and memory are connected by one or more high-speed buses. Loosely coupled systems, sometimes called multicomputers, have local memory for each processor—though like multiprocessors they sometimes have global memory for shared data.

In bus-based systems one recent event of note was the introduction of the long awaited Multimax computer system from Encore Computer Corp. of Marlborough, Mass. The Multimax has a 100-megabyte-per-second bus and can be expanded from 2 to 20 microprocessors with 4 to 32 megabytes of common memory. Encore's processors are based on National Semiconductor's NS32032 microprocessors, for which Encore's engineers have devised a system that expands the addressing from 24 bits to 32. On each processor card in the system are two processors, which share a 32 kilobyte cache memory based on 45-nanosecond static RAM (random access memory).

Another machine in the same general class is the Sequent system, introduced in September of 1984, which can be expanded from two to 12 processors and up to 28 megabytes of main memory. It has a 26.7 megabyte-per-second bus; the processing units are based on National Semiconductor's NS32032 and NS32332.

2.3 SOFTWARE

2.3.1 Introduction

The most important fact one must recognize in exploring the possibilities and constraints on further software development is the existence of an enormous base of applications, written in many different languages and running on many different machines. We know that this software has its limitations: it is often unreliable, much of it has never been fully tested, and most important, a majority of applications are mutually incompatible as they are presently implemented, particularly in the area of the user interface.

We have a vested interest in these billions of lines of code, and the billions of dollars worth of data the code manipulates. We cannot afford to abandon it, and the task of transforming it into something more workable in an open systems environment is monumental. All we can hope for is a gradual movement toward more workmanlike procedures in software development, and the slow emergence of more reliable, verifiable, usable, and mutually compatible applications.

Chapter 5 discusses some of the ways in which the present body of software can be improved for use in a restricted kind of open systems environment. The degree of possible improvement, however, is extremely limited, and we must look for a general solution only in the longer term.

For the longer term we offer two general approaches, or rather two examples of the many possible approaches. One of these is standardization. Here Ada has been selected for two reasons. First, it is an attempt at a comprehensive compiler; many people and organizations have contributed to its specification. Second, because Ada has been sponsored by the U.S. Department of Defence, it stands a better than usual chance of gaining general acceptance. Even Ada, however, has its drawbacks: because it is comprehensive (and because no subsets are permissible) it is cumbersome and difficult to use. Moreover, it cannot be entirely portable, and the difficulties of making it run efficiently on any particular machine are considerable.

Another hope lies in automatic program generators. These are perhaps more promising in the longer term because they remove one level of complexity from the concern of the designer. The program designer need only specify what his application is to do; the program generator interprets this specification in terms of the particular machine on which it is to be implemented.

2.3.2 Ada Compilers

Ada is a relatively new programming language which has introduced a number of concepts of great interest to the designers of open systems. The requirements for Ada were developed by the U.S. Department of Defence, under the leadership of David Fisher. Fisher consulted experts from the military, from industry and the universities, and not only in the United States but around the world. It was the first time a completely separate group defined the requirements for a programming language [Booch 1982]. The development of the language requirements took nearly three years. The Strawman documents appeared in 1975, Woodenman in 1975, Tinman in 1976, and Ironman in 1977. These documents stated in progressively greater detail what requirements the language was expected to satisfy.

Ironman stated that the key objectives of the language were to improve the reliability, readability, and maintainability of programs, along with more classical goals such as the improvement of portability and efficiency, declaring [Ironman 1977] that "the language should promote ease of maintenance. It should emphasize program readability over writability. That is it should emphasize the clarity, understandability, and modifiability of programs over programming ease." What was new in Ada was that for many years the main goal had been on creating languages that make it possible to write programs very quickly (e.g., Basic, RPG, COBOL). It took 20 years for computer scientists to recognize that this was not an appropriate goal.

While the development of a large system may take less than 5 years, the same system may have to be maintained for more than 20 years (e.g., OS, DBMS). Even if it takes a little longer to write it, it is far more important that a program be readable than writable. For 20 years after the system is written, programmers will have to read it in order to maintain it: the easier a program is to read, the easier it will be to maintain.

One important aspect of the success of Ada in open systems is the acceptance the language has achieved among the computer vendors. In 1984, DOD was struggling to get validated Ada compilers to market. In late 1985, the initial trickle has become a flood. Vendors are rushing to deliver machines that run Ada.

Furthermore, Ada run-time environments will soon support the partitioning of tasks across several processors in a multiprocessor system. Hardware manufacturers such as DEC and SUN seem to be jumping on the Ada bandwagon. Verdix, a Chantilly VA software house, has recently added a number of multiprocessor manufacturers (Sequent, Tolerant Systems, Flexible Computer, etc.) to the list of hardware families running the Verdix Ada compiler and the Verdix Ada development system. Also, the adoption of Ada by the DOD will ensure the development of Ada on machines from major computer manufacturers such as IBM.

Five important features of Ada in an open systems environment are described below:

1. *Standardization*. Ada has been standardized at an early stage in its development. It represents the first instance in the short history of programming languages of things being done in the "right order." First the specification was formulated, and then the design (i.e., the major architectural lines of the language). Next, the language was standardized—the ANSI standard was issued in February 1983. Great pains were taken to ensure that the description of the language was precise, and that everybody agreed on the definitions. Then a validation facility was produced: a tool to ascertain that compilers conform to the standard. Finally, the compilers appeared in late 1985.

2. *Package structure* is probably the most important feature of Ada. It creates a very clean separation between the visible part (i.e., the interface with the user) and the package body (the domain of the implementer).

The user's view of Ada facilities is defined by the *package specification,* which is the visible part of the package, (i.e., abstraction). The programs that provide their mechanism are called the *package body* (i.e., detailed implementation). It is a general rule in Ada that the user view (specification) and the implementation are always presented separately.

In the open system environment the package structure is plainly attractive. The system team and application team agree on the different packages that the system team must provide to the application team, and then they must agree on the application interface to each package. Each team goes away and develops the package bodies.

3. *Portability*. There is a commonly held view that to ensure a completely portable compiler one need only state the language fully, and then enforce it with standards and a validating compiler. However, Nissen [Nissen et al. 1983] argues strongly against this assumption. Ada is a case in point.

Since Ada has an adequate language definition from the point of view of compiler writers, it might seem reasonable to conclude that compliance with this definition is the only language-related issue to be addressed in specifying an Ada compiler. Such a conclusion is unwarranted for two reasons: inclusion of machine specific features, and compromises between total portability of Ada programs and efficiency of compiled programs. By total portability Nissen means the ideal situation in which every valid program has identical effects on all valid implementations in the open system environment.

Total portability is unfortunately not achievable. For instance, because floating-point hardware systems vary, programs will behave differently in different systems. A particular application program may have to use machine-specific features such as interrupts vs polling or ways of packing data specific to particular machines. Thus, the availability of a complete Ada standard still makes it necessary to specify machine-dependent characteristics of the language as implemented by a particular Ada compiler.

Further, Nissen [Nissen et al. 1982], in the study "Guidelines for the Portability of Ada Programs" has stated that the compromises made between the complete, machine-independent specification of Ada and its underspecification for efficient run-time implementation on many different systems make portable Ada programming a feasible but complicated task. Much of this complexity is not unique to Ada but seems inherent in contemporary programming languages where efficient implementation is an essential requirement for acceptance by users [Wallis 1982]. Ada is likely to be used in areas where efficiency of the running program is vital to the application (e.g., on-line transaction processing and real-time defence applications). An unacceptably inefficient implementation of Ada could result in the use of Assembler, with all the attendant problems whose identification by DOD sparked the initiation of the Ada development.

4. *Specification*. Closely related to the problems of portable Ada programming are the issues which arise in specifying the implementation of the language by a given Ada compiler. For example, the Ada type "integer" does not have any language-defined range; its range depends on the specific target computer. However, many computers can easily handle more than one integer size, and therefore do not define the range of "integer."

5. *Retargeting*. In an open system environment it is necessary to deal with the problems of retargeting, the process whereby, for example, an Ada compiler which currently generates object programs to run on one or more target computers is made to generate code to run on another kind of computer. Further, such code must be executed on host machines with different programming support environments or different operating systems from the original.

Although retargeting could potentially involve rewriting large parts of a compiler, most Ada compilers now consist of two parts, one machine-independent and the other machine-dependent. The latter must be rewritten to generate object code for the new target computer—like the Kernel in UNIX.

In addition, a run-time system must be constructed for the target. The retargeting of Ada compilers was considered sufficiently important by the DOD to warrant detailed consideration in the preparation of the Ada specifications. This emphasis is in stark contrast with similar exercises for previous languages such as COBOL, Fortran, and PL/1. This difference is partly due to the extensive use of cross-compilation in the open system environment, and the small part hardware manufacturers are expected to play in the Ada compilers market.

Retargeting issues considered in the specification include the availability of an Ada standard kernel, and specific questions concerning the production of optimizers for particular machines. A retargetable compiler should do the maximum possible amount of target-independent optimization. Such features as the object-code listing should be portable in all but the details of the actual code. Such portability maintains consistency of the user interface, which is important when programmers must work on several different machines in the open environment.

2.3.3 Automatic Programming

Automatic programming involves the application of the techniques of artificial intelligence to the general goals of automatic program construction and program transformation. To achieve these goals it is necessary to incorporate two kinds of knowledge. The first is knowledge of programming principles, including the programming language that will be used, and of the hardware and operating system on which the program will run. The second is knowledge of a specific application domain. Application domain knowledge enables the system to make use of constraints on the resulting program imposed by the application and help the system choose the most efficient implementation methods for a particular problem.

Automatic programming comprises a number of different disciplines such as:

1. Program synthesis, the aim of which is to construct programs from rigorous and non-algorithmic specifications describing what the program should do.
2. Program verification, which proves the correctness of programs by using mathematical techniques (such as theorem proving algorithms) to show that the programs behave according to their formal specification.
3. Optimizing compilation: optimizing compilers make the machine-executable code more efficient by using search techniques to find patterns in the use of variables, and a knowledge base used for storing useful segments of code.

Rich [1984] and Balzer [Balzer et al. 1983] summarize the aims of automatic programming as follows:

1. To eliminate the errors and loss of information that occur in creating, documenting and maintaining software.
2. To make it possible for non-programmers to use the computer to solve a problem (automatic code generation as one aspect of the task).
3. To make it possible for a programmer to express the purpose of the program in specifications that are close to the way he thinks, without having to make them computable.
4. To permit program maintenance at a level above the source code level.
5. To make possible a library of reusable software, a library of specifications, rather than of implementations which have a low potential for reuse.

Approaches to automatic program synthesis reported in the literature include [Scown 1985, and Waters 1982]:

1. Synthesis based on theorem proving. Programs are built in terms of mathematical theorems and verifying each step by proof (limited real world applications).
2. Deduction-driven synthesis based on transformation rules. The system develops a program from the user's specifications by means of fully defined transformation rules.
3. Knowledge-based synthesis. The system selects the programming techniques encoded in the knowledge base that most efficiently accomplish the specification of the algorithms and maps them into a total description of the program that is desired, and then codes the algorithms in the target computer language. (The correctness of programs generated by this method cannot always be guaranteed.)
4. Problem-reduction based synthesis. In this method the synthesizer decomposes the original problem into sub-problem specifications, which are assumed to be simpler, until elementary specifications are obtained.

Successful applications using automatic program include [Winston 1984]:

1. USE.IT (by Higher Order Software, Inc. Cambridge, MA), which runs on VAX superminicomputers and on IBM mainframes, creates correct Fortran and Pascal programs. USE.IT applies mathematical theorems that determine whether an error has been made. In addition to automatically generated code, USE.IT provides automated documentation. The software engineer using this tool performs more design work and less coding drudgery than usual.
2. CHi (developed at Kestrel Institute, Palo Alto, CA) is an example of a deduction driven system based on transformation rules. It is a prototype of a highly integrated intelligent programming environment. CHi selects the appropriate transformation rules to implement the user's specifications, producing lower-level code, like USE.IT.
3. GIST (developed at the University of Southern California Information Science Institute, Los Angeles, CA) is another example of a deductive transformation rule system. GIST's library of rules transforms a fairly natural human way of stating a problem into more executable statements.
4. The Programmer's Apprentice (PA) is under development at the Massachusetts Institute of Technology. On completion, the PA will allow a person to build a program from "cliches," fragments

corresponding to common algorithms represented in the system by "plans." These plans represent only the essential features of algorithms. Several thousand plans will be required to cover the domain of fundamental programming cliches to a useful extent.

In addition, the PA will be equipped with some knowledge of the problem. In using the PA, the programmer's job is to maintain an overall view of what must be done to accomplish the goal of the program. The editor's job is to keep track of the details of the implementation. The advantage of using the PA is that it eliminates errors for the system and the programmer because the programmer works not on code but on the plans; only the coding module of the system must work at the level of code.

5. BIS (developed at Artificial Intelligence Project, Milan Polytechnic University) utilizes the problem reduction approach. In BIS, during the top-down phase, program specifications are decomposed until primitive specifications are reached whose solving pieces of code are known. In the bottom-up phase, the primitive program segments are assembled to build up the final program.

2.3.4 Operating Systems

The purpose of an open system environment is to provide mechanisms which allow otherwise autonomous processors to cooperate in a consistent and stable manner, rather than to provide centralized control over all units of the environment. Distributed algorithms which allow cooperation among the processors in order to increase reliability, availability, and performance are employed in such an environment.

Today almost all computer vendors sell their own proprietary operating systems, written in languages dependent on their own machines. Every time a new machine is developed, therefore, a new operating system and new compilers must also be developed and expanded. Existing application programs must then be modified to run under the new operating system on the new machine.

Successful implementation of an open system environment depends on two important factors pertaining to operating systems:

1. The ability to transport a portable operating system to any machine in the open system environment.
2. An operating system in which some or all of the capabilities can be distributed.

A layered approach including the following steps can facilitate the development of a portable operating system:

1. As much as possible of the operating system and the compilers are written in a high-level language.
2. All machine-dependent parts of the operating system and compilers are isolated and written in a lower-level language.
3. The code generator for this lower-level language is then rewritten to generate machine instructions for the target machine.
4. The machine-dependent layer of the operating system is also rewritten for the target machine.
5. Finally, the entire operating system is recompiled by the modified compiler, to produce an operating system that will execute on the target machine.

A second factor in creation of a successful open system environment is dependent on the functional capabilities of the operating system. The algorithm used to schedule tasks will have a major impact on the perceived performance of the open system environment. The algorithm must be able to adapt to changing network conditions and processing loads at the various nodes.

One of the most important mechanisms which a distributed operating system can provide is the ability to move processes from one processor to another. This mechanism is called "process migration." Ni [Ni et al. 1984] proposes a distributed, user-transparent process migration algorithm for load balancing which is network topology independent. The algorithm works a compromise between the competing objectives of process-migration protocols: to maximize the utilization of resources while at the same time minimizing communication costs.

Issues pertaining to the implementation of UNIX in a distributed system have been much discussed recently. Panzieri and Randell [1985] describe the design and implementation of a network interface for use within UNIX user programs. They consider a number of geographically dispersed UNIX systems running on machines interconnected by a wide variety of data communication facilities such as packet switching networks, local area networks, radio links, and asynchronous lines. Panzieri and Randell suggest that the UNIX interface to a data communications network should be homogeneous, and as well-defined as that of ordinary UNIX I/O devices.

They propose interfaces which provide the abstraction of (possibly) very large datagrams, using a simple standard network addressing scheme which consists of <host number, port number> and do not in general hide the fact that a machine is connected to several data communications networks. In line with their aim of providing a basic standard uniform interface, they regard the fundamental task of the networking primitives as the transfer of a set of bytes between ports. When such transfer occurs between heterogenous processors, various other problems must be dealt with, particularly ones due to incompatible character-sets, byte ordering, and floating-point number representations.

A number of different approaches to the provision of UNIX interfaces to data communication networks have been reported in the literature by other groups. For example, University College London (UCL) has developed a UNIX program interface to networks that is very similar in its objectives to the uniform interface proposed by Panzieri. The UCL interface, termed "clean and simple" [Braden et al. 1983], provides UNIX programs with a uniform-access mechanism to various networks, including the British PSS and SERCNET and the DARPA Internet, and with the ability to transfer unbounded streams of bytes over these networks. The "clean and simple" interface is connection-oriented, and therefore supports applications such as remote terminal access.

Other relevant approaches are the Xerox Pup internet architecture, and the networking facilities of the Berkeley 4.3 BSD version of the UNIX system [UNIX 4.3 BSD].

The Pup architecture [Boggs et al. 1980] is based on the internet datagram protocol (IDP). This protocol implements a process-to-process datagram service among processes distributed over one or more different interconnected networks, and maintains a uniform-access interface to these networks. The IDP interface can be characterized by:

1. Primitive operations for transmitting and receiving "Internet Packets" of 512-byte maximum length across the networks.
2. A uniform datagram addressing convention, which applies to any network to which a machine is connected.
3. An interface which maintains the abstraction of well known addresses which can be statically allocated to specific servers.

The Berkeley 4.3 BSD [UNIX 4.3 BSD 1986] operating system supports a rather sophisticated interface for interprocess communication, both within a single 4.3 BSD system and among systems distributed over data communication networks. This interface (layer) maintains the abstraction of several address families (or distinct domains). A domain represents a communications environment characterized by specific communication facilities (e.g., the UNIX environment facilities) or by DARPA internet environments identified by virtual circuit facilities. These facilities are made available to user processes by means of a standard set of fourteen communication primitives. These primitives can be used for communicating in each domain.

Processes within a domain can communicate by sending and receiving messages between communications end-points, termed "sockets." A socket is a typed operating system object used by user processes willing to communicate in a domain. The socket-type (such as datagram, stream, and sequenced) is selected by the user process and determines the semantics of the communications that the user process will use. In the UNIX domain, for example, a stream socket will provide pipe-like communication, in the internet domain it will provide transport service-like communication instead.

At the user level, the abstraction of a domain may appear similar to that of virtual address. However, the abstraction of a domain does not hide the nature of the underlying communication environment. In order to communicate within a domain, processes may require knowledge of some of the characteristics of the communication environment that the domain represents: for example, in order to select an appropriate socket-type.

It is perhaps worth summarizing the basic structuring ideas for the distributed operating system under UNIX:

1. Interprocess communication and networking are separable issues. Hence, the interface communication driver is patterned on the standard UNIX I/O interface driver rather than on some existing or new interprocess communications scheme.
2. The actual networking protocols used should not be evident at the UNIX user programming interface (i.e., network protocols should be transparent to the user). The implication is that the user program should be presented with the abstraction of a datagram which can be essentially any size, from zero bytes upward.
3. As far as possible, user programs should be shielded from the fact that network addresses can take various forms.
4. Internetwork addressing should not be part of the network interface. It should be possible for internetworking to be implemented at the user program level, or the network driver level, or indeed both.
5. The actual network protocol should be implemented on the AT&T UNIX V.3.
6. The extended UNIX should provide a suitable basis for the programming of a variety of transaction-oriented network applications. The reason is that, in effect, all that the connection requires is a means of transferring bytes from one bit to another.

2.3.4 Distributed Database Management Systems

Distributed database management systems are only a small part of the overall application areas possible within the open system environment; it is, however, one of the most widely discussed. A distributed database management system allows data files to be distributed and managed on the network of geographically dispersed computers which constitutes the open system environment. The users of a distributed database management system can access the data as if it were all located at one site, since the actual distribution of data across the system is transparent to the application user. Because data in a distributed environment may be spread across many different computer sites, movement of data over the communication paths among the computers can cause a significant delay. Efficient file maintenance and query-processing mechanisms are required in order to limit such delays.

Research in distributed database systems has been motivated by a desire for improved reliability and better performance, and reinforced by the availability of inexpensive communications services and computer hardware.

Concurrent transaction processing is a means of optimizing system performance and resource utilization in such systems. However, unconstrained concurrent transaction processing can lead to undesirable situations such as loss of updates, nonrepeatable reads, and inconsistencies in the database. The problem of preserving consistency is more complex in a distributed database system because of, among other reasons, data replication and inherent communication delays. (See Chapter 6 for more detail).

The reliability of a distributed database management system depends on many factors, but notably on the individual reliability of each of its communications links and computing elements (nodes), on the consistency of the database, and on the distribution of its resources, such as programs and data files.

An increasing number of systems rely on wide-area communication networks for connectivity, rather than on dedicated links. The response-time perceived by each user of a distributed database management system is the sum of three components: the processing time at the host computer, communication delays in both the forward and return paths of the subnetwork (assuming the user is accessing a remote database), and the queuing delays at each node.

2.3.5 Commercial Distributed Database Systems

There are at present no truly distributed database management systems on the market. One which comes as close as any to the definitions offered here and in Chapter 7 is the Encompass system offered by Tandem. On the other hand, Encompass does not support vertical fragmentation or data replication: the user must explicitly update different copies of the same file. No definition of data fragmentation and allocation is incorporated in the main data dictionary (catalog) for the system.

The terms in which Encompass is discussed here have not for the most part been introduced in the present chapter; they are defined in Chapter 7.

In the Tandem implementation of Encompass the computational structure is based on a requester-server approach. In a requester-server structure communication between processes is by messages. Other features of this structure are:

1. A requester process may be multi-threaded.
2. A server process is always single-threaded.
3. A server process should be context-free—it should not be required to remember past requests.
4. Server and requester processes can reside at different sites on the network.
5. A single server process can perform its service on behalf of a number of different requesters.

Two-Phase Locking is used in Encompass for concurrency control, thus ensuring the ''atomicity'' and ''serializability'' (see the ''ACID'' test in Chapter 7) of each transaction. A lock can be obtained at either the record or the file level. No mechanism is provided for detecting deadlocks; applications can recover from deadlocks by means of timeouts.

Finally, in Encompass the data distribution mechanism has been designed to satisfy two main requirements: continuous availability and the autonomy of all sites.

The following commercially available systems embody at least some of the defining characteristics of a distributed DBMS:

1. IMS and SQL/DB, with CICS/ISC, by IBM
2. Datacom/DB with D-Net, by Applied Data Research
3. IDMS-DDS, by Cullinane
4. IDMS-DDBS, by International Computers
5. VDN, by Nixdorf
6. UDS-D, by Siemens
7. ADABAS/Net-Work, by Software AG
8. INGRES Star, by Relational Technology

The salient features of these systems are listed in [Ceri and Pelagatti 1984]:

1. *Foundation*. All these systems except VDN have been developed as extensions of existing centralized database management systems.

2. *Data model*. IDMS and UDS belong to the CODASYL class; IMS uses a hierarchical model; Datacom/DB, ADABAS, INGRES Star, and Encompass are relational, with inverted lists; VDN is a compromise between the CODASYL and relational models.
3. *Unit of distribution*. The base unit of distribution in all these systems is the file; only VDN, INGRES/Star, and Encompass allow the partitioning of files into smaller units.
4. *Data replication*. Five systems (D-Net, VDN, INGRES/Star, Encompass, and Net-Work) allow the duplication of data across various nodes in the distributed environment. D-Net and INGRES Star keep the maintenance of data invisible to the end user. In the other three systems only the master file is updated consistently; copies can be used only in performing "dirty" reads in transactions with no requirement for consistency.
5. *Distributed transactions*. Several of these systems already support Two-Phase Commitment, and it is easy to foresee that distributed transaction recovery protocols will be incorporated in all systems which claim to manage a distributed database. This is not the case in the more critical areas of replication, robustness, and concurrency control.
6. *Location transparency*. IDMS-DDS and INGRES Star allow users and application developers to view all the data within an organization as a single database. Users need not know where data is located in order to gain access to it; developers need not change their programs when data is moved or incorporated in new systems. These two systems embody the three main elements of location transparency: distributed access to data, a distributed data dictionary specifying the location and availability of data, and a distributed transaction processor to ensure that data consistency is maintained across the entire distributed DBMS. Other systems offer one or two of these elements, but not all three.

Table 2.4 is a summary of the features of research prototype distributed database management systems.

2.4 COMMUNICATIONS

2.4.1 ISDN: Overview

The concept of an integrated services digital network (ISDN) is simple: universal interfaces for digital communications applicable equally from the local-area network (LAN) level to the international wide-area network (WAN) level. Universal standards would make it possible to offer new

Table 2.4. Comparison of Features of Homogenous Research Prototypes.

	SDD-I	R*	DDM	D-INGRES	POREL	SIRIUS-DELTA
1. Fragmentation supported?	—	N	—	—	—	—
—Horizontal	Y	—	Y	Y	Y	Y
—Primary	Y	—	Y	Y	Y	Y
—Derived	N	—	Y	N	N	Y
—Vertical	Y	—	N	N	N	Y
—Mixed	Y	—	N	N	N	Y
2. Fragmentation transparency supported?	Y	N (allocation transparency)	Y	Y	Y	Y
3. Relational data model used?	Y	Y	N (Daplex)	Y	Y	Y
4. Interfaces provided						
—Query	Y	Y	Y	Y	Y	Y
—Host-programming language imbedding	N	Y (PL/I)	Y (Ada)	Y (C)	Y (Pascal)	N
—Precompiled applications	N	Y	Y	N	Y	Y
5. Query processing						
—With an initial allocation-independent analysis	N	N	Y	N	Y	N
—Separating global from local optimization?	Y	N	N	N	Y	Y
—Taking only transmission costs into account?	Y	N	N	N	Y	Y
—Using semi-joins?	Y	N	Y	N	N	N

6. 2-phase-commitment used?	N (4-phase)	Y	N (4-phase)	Y	Y	Y
7. 2-phase-locking used?	N (timestamps)	Y	Y	Y	Y	Y
8. Deadlocks are avoided (A) or detected (D)? How?	A timestamp-based c.c	D distributed detection	D distributed detection	D centralized detection	A preclaiming of locks	A prevention
9. Replication supported? —update method for copies	Y write-all	N —	Y write-all (regular copies only)	Y deferred updates	Y deferred updates	Y write-all
10. Requires global status of network sites?	Y	N	Y	Y	Y	N
11. Centralized (C) or hierarchical (H) computational structure	C	H	H	C	C	C

SOURCE: Stefano Ceri and Guiseppe Pelegatti. Distributed database. Principles and systems. McGraw-Hill Inc. 1984 p. 357.

services and provide flexibility to accommodate new types of equipment and application in the future.

At the moment, ISDN exists as a set of interconnection standards sanctioned by international standards bodies—a set of standards neither fully formed nor fully accepted. However, there has been enough work by authorities in enough countries (e.g., U.S.A., Japan, and European) and by enough equipment suppliers (e.g., AT&T, IBM, DEC, etc.) to convince even the sceptical that ISDN is real—or shortly will be.

The ISDN standards are published by the Consultative Committee for International Telephone and Telegraph (CCITT); they are called the "I-series" of recommendations. The heart of ISDN resides in the I-1.400 subseries, which defines the basic and primary access rate interfaces for layers 1, 2, and 3 of the ISDN architecture—corresponding to the same layers of the open system interconnection (OSI) model, also published by the CCITT. The other sets of I-series recommendations are:

1. I-1.100: the general concept of ISDN, terminology, methods and structure of the recommendation
2. I-1.200: the service aspects of ISDN, including bearer (carrier) services and teleservices
3. I-1.300: particular network aspects, including functional attributes, performance characteristics, numbering, and addressing
4. I-1.500: internetwork interfaces
5. I-1.600: maintenance principals and guidelines for user testing

For current users, it probably makes sense to think of ISDN as a substitute for existing analog voice-grade lines. In field trials, for example, customers have tended to use ISDN to keep down the cost of moves and changes of telephones and to allow easy institution of data lines. In the long run, however, ISDN will offer the opportunity of unique services, provided both by information management groups within organizations and by external vendors. ISDN should ultimately support transmission over the same wire-pair of everything from telex, telegraph, and radio paging to telephony, music, facsimile, hologram transmission, and slow-scan video. If ISDN is there, and the price is right, users will take advantage of it, particularly those in dynamic networking environments.

2.4.2 Problems with the ISDN Concept

1. The series of ISDN *standards* is not complete, in part because of major differences between international and U.S. requirements. Users must therefore be wary of variations in features implemented

or stressed by individual vendors. However, the lack of a single dominant international carrier makes it easier for CCITT recommendations to gain wide acceptance.

2. Common sense suggests that the installed base of telephones and keyboards might generate a *bandwidth requirement* at least 10 times that of the bandwidth required to transmit data in 1990 (see Table 2.5). ISDN itself may generate more data traffic. However, it is unlikely that new services alone will generate much new data traffic sooner than 1995. The installation of extensive fiber-optic networks in North America, Japan, and Europe has led some observers to predict a glut of transmission capacity by the end of the decade.
3. A great many vendors will have to support ISDN standards. Existing equipment will have to be upgraded. VLSI chips will not be available until final standards are set. On the other hand, international experience has created at least a handful of vendors with ISDN experience and equipment (e.g., Motorola Inc., Intel etc.).
4. Lack of vendor support, incomplete adherence to standards, and the need for expensive gateways between proprietary systems (e.g., IBM's SNA and DEC's DECnet) and the ISDN standardized system could be continuing problems.
5. Local loop lengths in the U.S.A. make ISDN transmission impossible in some areas; a limited bandwidth (144 kbps and 1.54 mbps) will mean limitations to some types of traffic.

Table 2.5. Bandwidth Requirements by 1990.

Phone:	500 million instruments
	32 kbs instrument
	1 hour/day use
Terminals:	40 million instruments
	50% remote
	4800 bps average
	1/2 hour daytime use
PCs:	40 million instruments
	50% communicating
	2400 bps average
	1 hour/day use
	Other 5 million instruments communicating
	4800 bps
	1 hour daytime use
	Total expected bandwidth requirements in 1990:
Voice:	8×10^{15} bits/day
Data:	4×10^{14} bits/day

6. There is no real demand for applications which must use ISDN. ISDN will be initially justified in terms of cost savings, and will later evolve into an opportunity for new applications.
7. To implement ISDN, the central and toll offices of the carrier will have to use ISDN digital switches. At first glance this would seem to indicate slow going for ISDN. However, those digital switches will be installed in areas where they are most useful: dense metropolitan areas, which will typically hold the highest concentrations of likely ISDN users.
8. The ISDN application may clash with past investments. The real problem lies in the computer operating systems. For example, for IBM to get computers talking together at 1.5 mbps demands rewriting both CICS and MVS.

2.4.3 ISDN's Promise for the Future

1. ISDN will provide two voice or data circuits over a single wire-pair, plus one for signalling. This should provide cheaper circuits, and should make possible the dynamic routing of voice and data circuits. In the long run there will be a significant cost saving in cabling and network termination units.
2. The signalling channel (D channel) will provide low-speed packet switching; the B channels will make it possible to mix and switch data along with voice. The ISDN standard interface will eventually do away with need for modems.
3. Integrated ISDN facilities will support voice mail, electronic mail, combined voice/data mail, and simultaneous display of data during telephone calls, as well as applications in financial services, medicine, and desktop teleconferencing.
4. The signalling channel will support alarm reading, meter reading, patient monitoring, etc., in situations where traffic is only occasional and does not require a particularly high bandwidth.

2.4.4 How Will ISDN Evolve?

In European countries, the PTTs (postal, telephone, and telegraph authorities) will establish the implementation schedule, applications, price, and availability schedules. It may not be worth the effort (as it appears Videotex is not) but at least the PTTs have the power to bring ISDN into being in a coherent way, at least within single countries.

In the United States the local telephone companies and the interexchange carriers will start establishing their own ISDNs—the result of which will be "islands" of ISDNs. For instance, one set of network switches might belong to the telephone company (the central office) and another to the carriers. There is no authority in place to ensure that the two can talk to each other.

One area in which international contention may occur is signalling technology. Traditionally, signalling has been done in the same band as the telephone call, but in the 1970s AT&T began implementing common-channel interoffice signaling (CCIS), a proprietary technique. Upon divestiture, AT&T took that technique with it. ISDN calls for similar signalling, SS7, a direct outgrowth of the CCIS technology. This time the technique is to be open—part of the ISDN standard.

2.5 ARTIFICIAL INTELLIGENCE AND THE USER INTERFACE

2.5.1 Introduction

Our ability to contrive appropriate user interfaces will undoubtedly be one of the key factors in the success or failure of open system environments. Because such environments will make available such a broad range of services, both familiar and unfamiliar, their users may well become hopelessly lost in a morass of procedural detail, or weary of learning new ways of doing things. The provision of simple, intelligible, and uniform means of interface between human beings and the complexities of the open system environment is an absolute necessity.

Rather than describing all the types and techniques of user interface, we have decided here to concentrate on tools emerging from that one field of investigation which will surely affect most dramatically the design of future user interfaces.

Artificial intelligence (“AI”—sometimes also called machine intelligence or heuristic programming) is an emerging technology which has attracted considerable publicity. One simple way of describing AI is to say that it is the business of devising computer programs to make computers smarter.

The computer programs with which AI is concerned are primarily symbolic processes which involve complexity, uncertainty, and ambiguity. These processes are usually ones for which algorithmic solutions do not exist, and in which searching is required. AI is a way of dealing with types of problem solving and decision making that humans continually face in the “real” world. These forms of problem solving differ from scientific and engineering calculations in that the latter are primarily numeric in nature, and are based on well-known algorithms that produce satisfactory answers. AI programs, in contrast, must often deal with words and concepts rather than numbers, and sometimes do not guarantee a correct solution—some wrong answers are tolerable, just as in human problem-solving.

AI problem solving is usually guided by empirical rules—rules of thumb called ''heuristics''—which help constrain the search.

Another aspect of AI programs is the extensive use of ''domain knowledge.'' Such knowledge must be available for use when needed during the search. It is common in AI programs to separate this knowledge from the mechanism which controls the search; changes in knowledge therefore demand changes in the knowledge base only, not in the search mechanism. By contrast, domain knowledge and control information in conventional computer programs are difficult to modify, as changes made in one part of the program must be carefully examined for their implications for and impact on other parts of the program [Buchanan and Shortliffe 1985].

Table 2.6 draws a comparison between AI and conventional computer programs.

2.5.2 Basic Elements

Nilsson [1982], a pioneer in AI, characterizes the components of AI in terms of what he calls the ''onion'' model (see Figure 2.3). The inner ring depicts the basic elements from which the application shown in the outer ring is composed.

Heuristic Search One common way of representing problem solving in AI is as a search among alternative choices. It is possible to present the resulting search as a hierarchical structure (i.e., a tree). Initially, the methods used for searching trees were ''blind;'' orderly search approaches ensured only that no solution path would be tried more than once. However, for problems more complex than games such methods were inadequate. Rules of thumb (empirical rules) were needed to assist in the choice of the most likely branches, in order to narrow the search.

Table 2.6. A Comparison of AI and Conventional Programs.

ARTIFICIAL INTELLIGENCE	CONVENTIONAL COMPUTER PROGRAMMING
Primarily symbolic processes	Often primarily numeric processes
Heuristic search (solution steps implicit)	Algorithmic (solution steps explicit)
Control structure usually separate from domain knowledge	Information and control integrated
Usually easy to modify, update and enlarge	Difficult to modify
Some incorrect answers often tolerable	Correct answers required
Approximate answers usually acceptable	Best possible solution usually sought

SOURCE: William B. Gevarter, ''An Overview of Artificial Intelligence and Robotics,'' Vol. 1, Artificial Intelligence. NBSIR 83-2799, 1984.

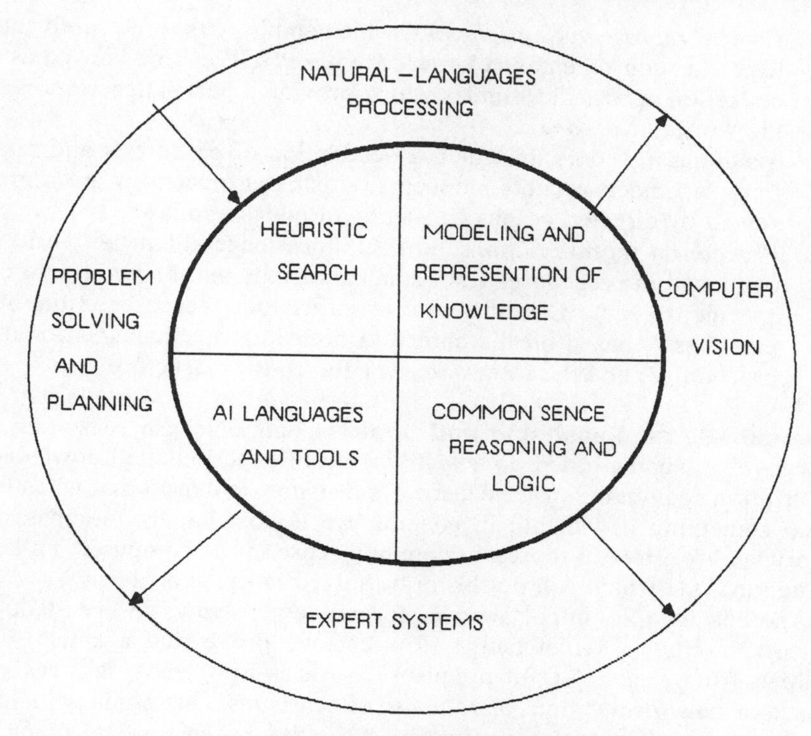

Figure 2.3. Elements of artificial intelligence.

Heuristic search techniques are one of the key contributions of AI to efficient problem solving. They operate by generating and testing intermediate states along a potential solution path.

Heuristic search techniques are now relatively mature and are codified in Barr and Feigenbaum's *Handbook of Artificial Intelligence* [1981].

Knowledge Representation The purpose of knowledge representation (KR) is to organize required information into a form readily accessible to an AI program, for use in making decisions, planning, recognizing objects and situations, analyzing scenes, drawing conclusions, and other cognitive functions. Thus knowledge representation is central to expert systems, computatorial vision, and natural language processing.

Representation schemes can be classified as either declarative or procedural. Declarative representations are of facts and assertions, while procedural representations are of actions. Declarative ''object oriented'' schemes can be further divided into relational and logical schemes. The principal knowledge representations are:

1. *Logical representation schemes,* which employ first-order predicate logic. In such schemes, a knowledge base (KB) can be viewed as a collection of logical formulas which provide a partial description of the world.
2. *Semantic networks,* in which a description of properties and relations of objects, events, concepts, situations or actions are represented by directed graphs consisting of nodes and labels.
3. *Procedural representations.* In PRs, knowledge about the world is contained in procedures (i.e., small programs that know how to do specific things). Classification of procedural representation approaches is based on the choice of activation mechanisms for the procedures, and the forms used for the control structure.

Common-Sense Reasoning and Logic Common-sense reasoning is low-level reasoning based on a vast amount of experimental knowledge. In acquiring common sense we learn, for instance, to expect that when we drop something it will fall; in general, we learn what to anticipate in everyday life. How to represent common sense in the computer is a key issue, and one which will not likely be solved in the short term.

Another field of importance is formal logic. How do we deduce something from a set of facts? How can we prove that a conclusion follows from a given set of premises? Logic is now enjoying a revival based on new formulations and the use of heuristics. One of many logical methods is called logical inference (i.e., the process of reaching a conclusion based on an initial set of propositions, the truths of which are known or assumed). This is normally done by "theorem proving:" a general automatic method for determining whether a hypothesized conclusion (theorem) follows from a given set of premises (axioms). First, using standard identities, the original premises and the conclusion to be proved are put into clausal form. The conclusion to be proved is then negated. New clauses are then automatically derived using resolution and other procedures. If a contradiction is reached, then the theorem is proved.

AI Languages and Tools AI programs tend to develop iteratively and incrementally. As the programs are evolutionary, creating AI programs requires an interactive environment with built-in aids such as dynamic memory allocation as the program evolves, rather than memory allocation in advance (as in most other types of programming). Research in AI programming has also revealed that expressing functions recursively (i.e., defining them in terms of themselves) leads to a great simplification in the writing of programs. AI programming languages therefore tend to support recursive processing. Finally, the main concern of AI programming is the manipulation of symbols, rather than numerical computation.

Two general AI languages, Lisp and Prolog, have evolved in answer to these programming requirements. Lisp has been the primary AI programming language; about 1960 John McCarthy at M.I.T. developed Lisp as a practical list-processing language with recursive-function capability for describing processes and problems. Prolog, a logic-based language, has appeared more recently (1973) and has gained favor in Europe and Japan. Prolog was developed at the University of Marseilles AI laboratory by A. Colmerauer and P. Roussel. Additional work on Prolog has been done at the University of Edinburgh in Great Britain [Colmerauer et al. 1981]. A number of AI languages have been developed as extensions of, improvements upon, or alternatives to Lisp. These include:

1. System Programming Languages (Lisp Level)
   - Pop-2 at University of Edinburgh 1976
   - Sail, at Stanford 1969
   - QA4 and QLisp at SRI 1968, 1972
2. Deduction/Theorem-Proving Languages
   - Planner and Microplanner at M.I.T., 1971
   - Conniver at M.I.T., 1972
   - Popler at U. of Edinburgh, 1972
   - Amord at M.I.T., 1977

Fahiman and Steele [1982] state that desirable features of an AI environment are:

1. Powerful, well-maintained, standardized languages.
2. Extensive libraries of code and domain knowledge (that is, a facility should support the exchange of code with other AI Research facilities).
3. Excellent graphic displays: high resolution, color, multiple windows, quick update; and the software to permit the easy use of these.
4. Good input devices.
5. Flexible, standardized inter-process communications.
6. Graceful, uniform user-interface software.
7. A good editor that can deal with a program on the basis of individual program structures.
8. Isolated machines are nearly useless. Good network interfaces, internal and external, are therefore essential.

2.5.3 Principal User-Interface Applications for AI

Natural Language Natural language interfaces (NLI's) were the first commercial AI product. Robot (now Intellect, by the Artificial Intelligence Corp., of Waltham, MA) in 1980 was the first into the market, and now runs in over 200 installations. Robot is intended for use as front-end interface to information retrieval application in areas such as finance, marketing, manufacturing, and personnel. Robot uses a formal grammar to parse the user's natural language query into an internal representation that sets off a database search. The system draws on knowledge of the structure and contents of the database, as well as a built in data dictionary, and an application-specific dictionary. When a query can be interpreted in more than a single way, Robot resolves the ambiguity by assigning preference rating to the different paraphrases, choosing the one which is rated most highly in terms of its consistency with information in the database. If necessary, Robot asks the user which of several interpretations is correct.

Another interface approach that emphasizes semantics has been taken by Roger C. Schank, founder of Cognitive Systems Inc., New Haven, Connecticut. Cognitive Systems products "map" natural language input into conceptual representations based on a theory (developed by Schank at Yale's Artificial Intelligence Laboratory) of conceptual dependencies, which capture the meaning of the input. In addition to storing the problem domain in various knowledge structures, such as scripts, these knowledge structures aid interpretation by providing the system with expectations, contexts in which the input is intelligible.

At the present time, individual natural language interfaces must be specialized for particular subject-matter contexts if they are to interpret words and phrases correctly. It should also be noted that although today's NL systems, and indeed AI systems in general, are limited in knowledge, the systems are usually not aware of these limits.

Several commercial machine translation systems are now available. System such as Logos, from Logos Corporation of Waltham, MA, is designed for business use. Logos works in partnership with a human translator. Another system, ALPS, from Automated Language Processing Systems Inc. of Provo, Utah, takes a slightly different tack, in that the translation process operates in an interactive rather than a batch mode. The interactive mode not only finds unknown words but also words with ambiguous meanings in context. The ALPS dictionary accommodates word strings or phrases as well as single words. Instead of one large dictionary ALPS utilizes several smaller reference dictionaries; it also builds a separate dictionary for each document. The document dictionary can be fine-tuned for the document's specific context without affecting dictionary definitions that will be applied to documents written in other contexts [TUCKER 1984].

Expert Systems Expert systems (also called "knowledge-based systems", KBSs) are most often used as intelligent assistants or consultants to human users. They can be used to solve routine problems, freeing the expert for more novel and interesting ones. Expert systems can bring expertise to locations where a human expert is unavailable or to situations in which an expert's service would be very expensive. Among other achievements, expert systems already have:

1. Acted as trouble-shooting advisers for oil well drilling operations;
2. Advised physicians on treatment for suspected bacterial infections in the blood; and
3. Configured complicated computer systems in a fraction of the time required by an experienced engineer.

Artificial intelligence reasoning and problem-solving techniques allow expert systems to draw conclusions that were not explicitly programmed into them. Expert systems use information which is not always entirely consistent or complete, manipulate it by symbolic reasoning methods without following a numerical model, and still produce satisfactory answers and useful approximations. Naturally, the more complete and correct the knowledge, the better the system's output. Some of the techniques and elements that make possible an expert systems novel inferences are knowledge acquisition, heuristics, knowledge representation methods, common-sense reasoning, and logic.

A problem lends itself to an expert system approach, according to Scown [1985], when:

1. A solution to the problem has to be a high payoff if it is to warrant the development of a system.
2. The problem can be solved only by an expert's knowledge, and not by utilizing a particular algorithm, which traditional programming could handle.
3. Access is available to a willing expert who can formalize the knowledge needed to solve the problem.
4. The problem doesn't necessarily have a unique answer. Expert systems work best for problems that have a number of acceptable solutions.
5. The problem changes rapidly or knowledge about a problem is constantly changing (as in the continuing discovery of causes and treatments of diseases) or solutions to problems change constantly.

The development of an expert system is generally not considered finished when the system is brought on-line. Development continues, with programmers updating the system's knowledge and processing methods to reflect progress or modifications in the problem area, of the system.

Because expert systems cannot yet learn on their own, they tend to require a continuing relationship with a human expert.

Expert systems are one of the most commercially rewarding areas of artificial intelligence implementation. The largest and most intensively used commercial expert system is Digital Equipment Corp XCON (Expert Configurer), formerly called R1 at Carnegie-Mellon University, where Prof. McDermott developed the initial 500-rule prototype. This system is used on a daily basis in Digital's worldwide manufacturing operation to configure VAX and PDP-11 computer systems ordered by customers. Dennis O'Connor, who spearheaded the XCON development effort, reports that as of the beginning of 1985, XCON utilizes over 4,200 rules to accomplish the task, taking less than two minutes to configure most systems. In addition to making changes or corrections to ensure that necessary pieces of hardware and other components are included, XCON checks orders for correct cable lengths, power requirements, and many technical details; it then prints reports, with several diagrams, to assist in assembling on the plant floor and at the customer's site.

Although expert systems can do a great many jobs, today they have limitations. With some exceptions, they cannot be used in situations in which real-time responses must be fast—say, for direct machine control. Today's expert systems are special-purpose, rather than general-purpose. They do not embody common sense: that is, they cannot use knowledge of the world to generalize successfully. In addition, though some can explain how they arrived at a conclusion, they cannot evaluate the validity of the process.

Chapter 3

The Architecture of Open Systems

The structure of an open system embodies many different technologies and relies on techniques from a number of different areas within the computing field. Clearly there are many different ways to utilize these techniques and the emerging technologies. This chapter will explore an architecture for open systems. Some components will be discussed in detail in this chapter, while others will only be introduced here and discussed in greater detail in subsequent chapters. Also, as illustrated in the preceding chapter, many areas of the industry are rapidly changing. This chapter will also examine the effect of emerging technologies on the architecture of open systems.

3.1 A LOGICAL ARCHITECTURE FOR OPEN SYSTEMS

A computer system can be viewed, at least abstractly, as a number of layers: a command language (or user shell), applications (including utilities and development tools), application services (e.g., file management, directories, print queues, etc.), system services (kernel operations, process communication, scheduling, etc.), and hardware. A more refined view can be found in Dijkstra [Dijkstra 1968] or Brown [Brown 1984].

A similar view can be adopted for open systems. Figure 3.1 presents a logical view of an open system. As with an operating system, the topmost level represents the *shell* or *user interface*. In the context of an open system, such an interface may be dependent upon the individual user and may differ on different systems. For example, a system administrator might have a different interface than a financial analyst, even though they may both be using the same system. Alternatively, the user of a word processor might be using a microcomputer and a programmer might be using a mainframe. Because interfaces differ greatly depending upon the application and system, they are not generally discussed within this book. However, some general principles in the design of user interfaces for applications are discussed in Chapter 5.

Beneath the interface layer lies the collection of *applications* available to the user. In this framework, a developer can also be considered to be a user making use of certain specialized development tools, e.g., compil-

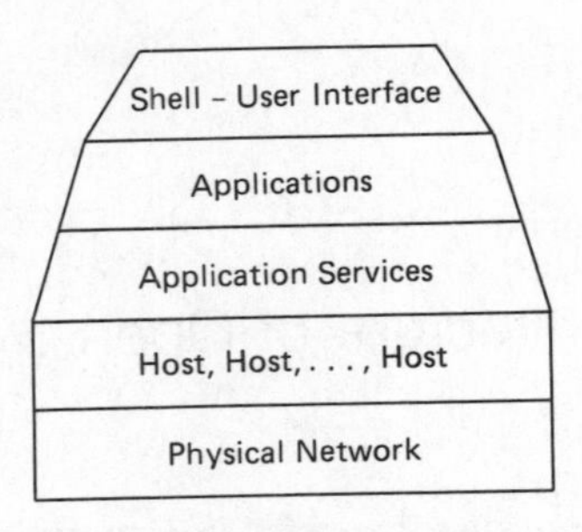

Figure 3.1. Logical architecture of an open system.

ers, editors. The applications, in turn, make use of a number of *application services*. These services provide mechanisms and means for ensuring and aiding the operation of applications within an Open System and across a number of different computer systems. Three classes of services are particularly important: *database management services, communication services,* and *directories*.

Database management services and the associated databases provide the basis for the distribution and management of data over multiple sites. These services also provide the means for user and/or application query of databases, local and remote, reliable transaction services used by applications, backup and recovery, etc. Databases in the context of an Open System are discussed in more detail in Chapter 8.

Communication services provide the means for electronic mail, routing, file transfer, remote process execution or logon, etc. It also provides the basic mechanisms for message transfer between applications at different sites. Because of the importance of data communications, Chapter 6 provides a detailed look at communications and associated services.

Finally, directories provide the means for identifying the entities within the system. In particular, they allow one to identify users and to name system-wide objects, specify their properties, location, associated privileges, ownership, etc. They can also be used to aid in the development, maintenance and operation of the entire system. Directories are discussed in more detail in Chapter 4.

Together this primary set of services provides the basis for additional application services, such as network resource management or performance analysis.

Application services, as described above, provide a layer which spans multiple hosts. Pragmatically, each application service is realized by a set of processes on each host within the entire system. Hence, the layer beneath application services, the *host* layer, must provide the foundation for all application services. In the presented view here, each host consists of both hardware and software, in particular the host operating system.

Beneath this layer one has, at least logically, the *physical network* layer, which can be viewed as the base connecting all hosts together; this is further discussed in Chapter 6.

It is clear that the host layer is very important: it binds applications and application services to the network. Given the view of rapidly changing technology outlined in the previous chapter, it is also an area of great concern. The remainder of this chapter is devoted to an examination of the role of a host, particular the operating system, in the context of the Open System architecture discussed above.

3.2 THE ROLE OF THE OPERATING SYSTEM IN OPEN SYSTEMS

Traditionally, an operating system is a control program responsible for a variety of tasks: resource management, scheduling, access control, device control, etc. In the context of an open system, these tasks do not change—the operating system is still required to control the facilities of the particular host.

In the architecture outlined in the previous section, however, an operating system for a host in an open system has an expanded set of roles centered on the application services.

3.2.1 Communications

Operating systems have always been responsible for interprocess communication. Typically, such interprocess communication is one-to-one: a sending process sends a message to a single receiving process and gets a response back. In an open system this role must expand, since processing and storage facilities may be divided among several hosts connected by a network. In this context, the interprocess communication may have to be expanded to allow for one-to-many communication [Cheriton and Zwaenepoel 1984, Boggs et al. 1980]. Such communication is useful when the identity of the desired receiver is not known, as when new hosts are added to the system. It may also provide a more efficient means of communication than the sending of separate messages, especially when the hardware provides efficient facilities for broadcasting.

The operating system must provide facilities for application services to achieve application-to-application communication. With reference to the ISO/OSI Model [Zimmerman 1980], the host operating system is responsible for the functions of Transport, Session, and Presentation layers. It must provide these services to applications and application services. In addition, network and database services may need access to such facilities as timers, time of day, process suspend/resume, all of which must be provided by the host system.

Furthermore, since applications and application services will be realized as a set of distributed programs, languages for implementing these will have to be able to communicate and synchronize activities. Such primitives must be provided by the host system [Natarajan 1985, Allchin et al. 1983, Hoare 1978]. These primitives may be supported by additional facilities for managing message queues, signalling a particular process on message arrival or transmission error [Leblanc et al. 1984]. Language constructs for these activities are discussed further in Chapter 5.

3.2.2 File Management

File management is often a large part of what an operating system must do. In an open system the file system is only partly resident on a single host and hence not under the direct control of a single operating system. This presents numerous problems, including data access, concurrency control, and deadlock control.

A number of different solutions to these problems have been explored. For example, data access can be provided in terms of various "units"—an entire file, a sequential subset of a file (Cambridge File Server [Dion 1980, Mitchell and Dion 1982, Needham and Herbert 1982]), a page (LOCUS [Popek et al. 1981, Walker et al. 1983]), or a subset of a page (XEROX Distributed File System (XDFS) [Mitchell and Dion 1982]).

Concurrency control provides a means for access to shared files. A common approach to concurrency control is the single-writer, multiple-reader policy. That is, a transaction can read a file if and only if the file is not being written, and a transaction can write a file if and only if the file is not being read or written. The unit of concurrency control, that is, the entity to be locked, can be an entire file or a page.

Deadlock control can also be approached from a variety of ways. For example, the FELIX file server [Fridrich and Oldor 1981] requires that a transaction declare all files before starting and hence detects possible deadlocks before a transaction begins. In the XDFS system, a transaction requests a page or pages and then waits for a certain amount of time before it times out. After a while it must retry acquiring access to the needed pages.

Other problems with file management which arise involve allocation of files (i.e., on which host a file should be stored), and access (e.g., can a user change hosts and still access all his original files). Work in this area has concentrated primarily on hosts having homogeneous operating systems, i.e., the same one on each system. In particular, a great deal of work has been done using UNIX, e.g., the Sun Network File System [Sandberg et al. 1985], LOCUS (Popek et al. 1981, Walker et al. 1983].

A collection of heterogenous hosts presents greater problems; some approaches to these types of problem are discussed in the following section.

3.2.3 Naming

Hosts in an open system must also deal with naming problems, related to the problems of file management. In the case of files, for example, a set of homogeneous hosts implies that each individual host has a similar file name space. This can be extended in a consistent fashion [e.g. Sandberg et al. 1985] to provide a network-wide file name space.

Other naming problems arise as well: naming of individual sites, naming of applications, processes. Again, a homogeneous environment simplifies the problems. In a heterogeneous environment, hosts may be required to map names from different hosts onto their own name space and vice-versa. Directories (Chapter 4) can be used to greatly simplify this particular set of problems.

3.2.4 Database Support

Concurrency control provides a necessary mechanism for controlling shared access to data. This is also very necessary for database management. In addition, host operating systems must also provide for reliable transactions for database access and update. This, at the least, requires a single protocol (e.g. two-phase locking protocol or a timestamp protocol, see Chapter 8) to be used throughout the hosts in a system.

3.3 COPING WITH HETEROGENEOUS OPERATING SYSTEMS

It is clear that a collection of heterogeneous computer systems poses a number of difficult problems. Nevertheless, it is desirable to include a variety of different types of systems in an open system.

3.3.1 Limited Set of Operating Systems

If certain problems can be more easily handled with a single operating system on all hosts, then, perhaps, dealing with only a small number of operating systems may not introduce too many complexities. This approach is similar to what a number of vendors have done. For example, Intel's Open Net network architecture accommodates host systems which can run either RMX or Xenix.

Digital Equipment's and IBM's approaches to networks of machines are similar. In each case, the vendor has developed a proprietary network protocol (DecNet and SNA, respectively) which forms the basis for connecting hosts. In principle, a network of the vendor's hosts, regardless of which of the vendor's operating systems is running on each of the hosts (at least within a certain family of vendor operating systems), can operate via the proprietary protocol. Hence, providing that one is willing to remain vendor specific, some of the problems of "different" operating systems can be avoided.

3.3.2 Portable Operating Systems

One obvious approach to coping with different host systems is to select a single operating system and then port it to each host one wishes to connect to the network. In essence, one is reducing the heterogeneous operating system problem to one of porting a chosen operating system. Various approaches for developing portable operating systems have been explored [Baskett et al. 1977, Powell 1977, Cheriton 1982].

Perseus [Zwaenepoel et al. 1984] was an interesting experiment in the development of a portable operating system for a distributed computing system involving different hosts (including a DecSystem-10, VAXes, and an IBM 4331). The goal was to build a general-purpose operating system that would be portable over a range of target machines. The operating system was implemented in an augmented version of PASCAL facilitating systems programming, exception handling, and separate compilation.

In principle, to do this one should look for an existing operating system or design an operating system which provides most of the capabilities required, is itself simple, well-structured and perhaps written in a high-level language. Such features make it easier to port the operating system to new hardware. Of the existing commercial operating systems, UNIX is the only viable candidate. It runs on both large and small machines and has been ported to a variety of hardware, is written in C, and it provides a number of basic facilities.

Much of early work on the development of operating systems in general focused on a structured approach for the development of actual operating systems [Haberman et al. 1976, Dijkstra 1971]. A common approach is to structure the operating system as a number of *virtual machines* (see Figure 3.2).

For each of the virtual machines, Machine k is "implemented" on Machine $k-1$. Machine 0 would correspond to the actual hardware, and Machine n to the user's view of the system. If the interfaces between the machines are clearly specified and well-defined, one need only implement Machine 1 on Machine 0 in order to "port" the operating system. This, of course, assumes that the other "machines" are implemented in a language which is portable, and that a compiler exists for Machine 0.

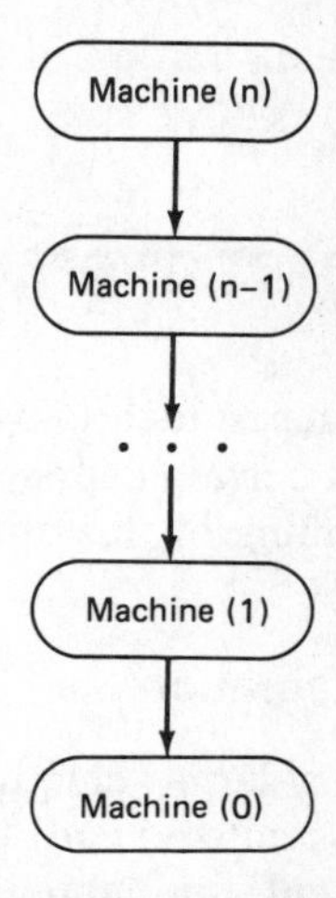

Figure 3.2. Virtual machine view.

This approach to the development of portable operating systems has been proposed in a variety of forms by a number of people: e.g. Dijkstra [1968], Goldberg [1972], Habermann et al. [1976], and Valdorf [1984]. It is also similar to the way in which UNIX is implemented: its "kernel facilities" correspond to Machine 1 and the remaining "machines" are defined in terms of C-library functions and programs.

3.3.3 Virtual Operating System

An alternative approach to a portable operating system is to define a virtual operating system. In such an approach, one specifies the set of system objects, properties, and functions belonging to a hypothetical operation system. These objects and functions are available to all applications and, moreover, are the only "operating system" functions and objects that the application may use. In turn, the virtual operating system is implemented as a layer on top of existing host operating systems.

In contrast to a portable operating system, one may choose to implement a virtual operating system for which Machine 0 of Figure 3.2 is an existing host operating system. Such an approach may not require as many virtual machines. Also, it may be less complex since (a) it may only require "high-level" operating system facilities (e.g., create file, allocate memory), and (b) it need only support applications and application services if one assumes the architecture of an open system as presented in Figure 3.1. Furthermore, the actual porting of this virtual machine may be simplified if each of the host machines supports a standard language with capabilities for the implementation of such a virtual machine (e.g. MODULA II, ADA).

3.4 TECHNOLOGICAL IMPACT OF OPEN SYSTEM ARCHITECTURE

In the light of the rapid changes in technology outlined in Chapter 2, this section considers the effect of certain changes on the architecture of open systems outlined at the beginning of this chapter.

3.4.1 Very Large Scale Integration

The increase in the use of very large scale integration will result in more powerful microcomputers, specialized hardware, and advanced computer architectures. Although it will undoubtedly cause changes in hosts, workstations, and other devices, it will also encourage the development of open systems.

First, creating hardware, whether it be new processors or mapping software, like compilers or communication translators, into silicon, demands that certain standards be set and adopted. Because the open system architecture attempts to make use of standards wherever possible, this will help stabilize certain areas. In the layered view of an open system, this is equivalent to various services migrating among levels: e.g., certain operating system or communication services be implemented in hardware rather than software.

Second, improved hardware technology will help to overcome some of the possible performance problems associated with the reliance on various standards, e.g., in communications. As standards become defined and accepted, technology will permit these standards to be "implemented" in hardware to gain performance. In some areas, like communications, this has begun already.

3.4.2 Advanced Computer Architectures

The emergence of advanced computing architectures, such as dataflow machines, parallel computers, supercomputers, knowledge machines, etc., will also pose many problems. On the one hand there will be the difficulty of integrating such machines into an open system, as in dealing with a host with an heterogenous operating system.

On the other hand, such machines offer the possibility of powerful and/or specialized application servers. The open system can make the resources of such computing devices available to a number of users or be incorporated into an open system without users' being explicitly aware of any change in hardware.

3.4.3 Advanced Devices

The advances in technology also hold the potential of different types of specialized peripheral devices, such as digitizers, and optical disks. Standardization of communication hardware and software can make such devices available to users of an open system. The incorporation of such devices into a system might be done in such a fashion that the user is unaware of their presence. For example, the use of an optical disk for database storage might replace a magnetic disk, yet its presence might be transparent to the user, and for that matter to the database services themselves.

Chapter 4

Directories and Dictionaries

Dictionaries and directories are both essential components of an open system. The data dictionary contains the fundamental definition of all the data resident at each site in the system. The database administration at each site controls its own dictionary, and can therefore maintain its autonomy while enabling users at other sites to gain access to its information. The directory is system-wide, and is controlled by the network administration; it shows users at all sites how to gain access to the resources or data they require. Although the directory plainly must at some level define the data at all sites, it does so in a general way; its main contribution is routing information.

When a user seeks information in an open system environment, his query is submitted first to the local system. If it is available locally, the search is conducted in terms of the definitions in the local directory at his own site. Only if what he requires cannot be found locally is the global directory invoked. A preliminary search is carried out in terms of the more general data definitions therein. If it is successful, the system-wide directory provides a procedure for gaining access to likely sources at one or more remote sites.

The issues pertinent to the implementation of dictionaries and directories are many and complex. Fortunately, almost all of them are aspects of the field of database management systems; the reader can find ample detail in DBMS textbooks. This chapter therefore concentrates on a problem which is specific to dictionaries and directories, and which is common to both: the problem of universal naming conventions.

Naming in a local dictionary can follow locally established conventions. In most cases local customs in identifying data will already be long established. To alter these to fit any universal scheme would be inconvenient and expensive; it would also violate the autonomy of individual members of the open system, who should know best how to identify their data. Data dictionaries are therefore normally the responsibility of the individual organizations which participate in the open system.

Naming conventions and access protocols in a system-wide directory must be universal, because the directory identifies all the data as elements of a single system, and because it is used by every site in the system. On the other hand, universal names are quite useless unless they can be "mapped" into the definitions in the local dictionaries. A considerable portion of this chapter is devoted to various schemes for universal naming, and to ways of moving from the local to the global and vice-versa.

The problems encountered here are more difficult than they might at first appear, because once he leaves his own site the user will have little notion how to proceed. He will probably not be able to ask for a file by name. He must be able to identify what he wants by less direct means, by content for example, just as he might browse in the subject index at the local library.

To ensure high availability and enhance performance it is sometimes necessary to replicate directories. Most access to directories is look-up, not update. Thus, in principle, multiple copies of a directory distributed around the network, permit many look-ups to be local, rather than involving network interaction and delay. Partitioning of catalogue information among directories also enhance performance due to locality. Terry [1984] analyzes a number of mechanisms for partitioning of information in name servers.

Effective administration of a distributed name domain is essential to a robust system. Sites should remain autonomous to the greatest extent possible for both technical (e.g., performance or availability in the face of partitioning) and non-technical (e.g., accounting or authentication) reasons. In particular, the failure of remote hosts should not prevent local clients from accessing directories that are stored locally.

4.1 SOME DEFINITIONS

Although most of the terms used in this chapter will be familiar to most readers, the meaning of some of them tends to vary from context to context. The following definitions are in force throughout the chapter.

Data Dictionary A collection of the names of all data elements used in a software system, together with relevant properties of those items: length of data element, representation, etc. [ANSI/IEEE 1983]

Name A linguistic object expressed in some language. Names correspond to entities in some universe of discourse—that is, a shared model of some part of a real or hypothetical world which is understood in some way by all participants. The correspondence between names (in the language) and entities (in the universe of discourse) is the relationship of denoting. A name denotes the entity to which it is bound.

Two distinct kinds of names can be identified:

1. *Primitive*. A name, assigned by a designated naming authority, the internal structure of which need not be understood or have

significance to users of the name. Its use is entirely internal to the system.
2. *Descriptive*. A name that identifies an object by means of a set of assertions concerning the properties of the object. It is used by the human users of the system.

Naming Authority A source of a central repository for names. The only constraints imposed upon naming authorities are that all of the names it provides must be:

1. Expressed in a prescribed language, and
2. Unique within the domain of the naming authority, which must never hand out the same name twice.

A further important distinction between the kinds of name stems from the fact that names can be used to identify sets of objects. This identification can be of three kinds:

1. A *generic* name is used when it is desirable to name a set of objects from which exactly *one* object is to be selected.
2. A *multicast* name is used when it is desirable to name a set of objects from which *all* objects from that set are to be selected.
3. *Synonyms* are two or more distinct names that identify the same object. A generic name or a multicast name can be a synonym.

Directory A directory is a collection of catalogue entries. With each directory is associated a particular name-prefix. A directory holds entries for all objects whose name consists of that prefix plus some path component. Within a directory names are of two kinds:

1. *Generic*. A generic name is used to indicate that the named object represents a set of equivalent names. That is, a generic name maps to a variable number of object names.
2. *Alias*. An alternative name. The use of aliases allows inverse mapping: that is, the mapping of any one of a set of names to a single object.

The principal aims of a directory are:

1. To insulate the user, or the user's application program, from the characteristics of the communications network.
2. To permit the movement of users and applications among sites, or among processors within a single site, without affecting other users who must communicate with them.

Agent A directory agent is a process which uses the directory system, accessing it on behalf of its associated user. The object entry for an agent must contain a globally unique agent identifier and a password to verify an authentication request. It is important to keep a list of the group of which the agent is a member.

Servers are recognized as a special type of agent and report additional information which reflects that special status. This information includes the various media access protocols by which the server may be accessed and the various object manipulation protocols understood by the server.

To contact a server, the application program must know the low-level (media) protocol used to transmit requests to the server, as well as the identifier referring to the server under that protocol.

In a directory the agent definition supplies global information and the server definition local information used in the coordination of activities.

Directory System A group of servers which interact with one another to provide all the elements of the system directory service to users.

4.2 DATA DICTIONARIES

The terms "data dictionary" and "catalog" are effectively synonymous in the present context—although there are significant differences between their meanings in the context of distributed databases. Although the major role of the dictionary is in management of the database, it has useful side-effects in other areas.

The data dictionary is a tool for the effective utilization of all objects in the database. It is a collection of definitions and information about such objects: a central repository of information that can be accessed by all areas of an organization.

A dictionary automatically provides data definitions for several database management systems such as IMS, TOTAL, ADABAS, and IDMS. In fact, several major software vendors have bundled their dictionary and the DBMS products; by this means the DBMS SCHEMA and SUBSCHEMA definitions can be produced only from the dictionary.

There are several advantages to generating data definitions from a dictionary [Leong Hong and Plagman 1982]:

1. It relieves the application developers from defining and coding data definitions; likewise, it relieves the database administrator of much of the detail work involved in the data definition language.
2. The dictionary is the source of all entities and attributes. Because all data definitions are automatically derived from the same source, we

reduce the inconsistencies and errors that result from manually prepared data descriptions.

3. The dictionary is the source of all data definitions. We can more carefully control changes that affect multiple data structures. If a change is made in a particular data element, the cross reference capability of the data dictionary will tell us every place this data element is used.
4. Dictionaries provide an audit trail of all changes made in data definitions. This audit trail includes the date and the initiator of these changes.
5. By centralizing the control and generation of data definitions, changes can be made only with the knowledge and consent of data administration. Before a particular data definition is changed, all users of this data definition can be notified of the impending change. Thus, all changes can be made in a controlled, scheduled manner.

Some of the secondary benefits are [Durrell 1985]:

1. The data dictionary can support structured analysis and design. It can be used to document data store, data flow, and process entity types. As such, it is an efficient way of portraying system design details to the user.
2. The dictionary significantly reduces the cost of program maintenance by enforcing consistency in naming and format variation.
3. By using the dictionary to generate the source program data definitions, the data portion of the program is actually mapped to the documentation. This direct mapping between documentation and system definition guarantees the accuracy and currency of the data documentation. After a system is implemented, all data changes would first be made in the dictionary. This assures that the data documentation will be kept up to date, and the data processing staff will certainly have more confidence in its accuracy.

4.2.1 Dictionary Management in a Distributed Database

The dictionaries of distributed databases store all the information which is useful to the system for accessing data correctly, efficiently and verifying that users have the appropriate access rights to them.

In distributed databases dictionaries are used for:

1. Mapping data referenced by applications at different levels of abstraction into a physical data-space.

2. Data allocation: the access methods available at each site, as well as statistical information, are required for producing access plans.
3. Verifying the access rights of users and application programs.

Dictionaries are usually updated when the users modify the data definition; for instance, when global data structures (relations) or fragments are created or moved, local access structures are modified, or authorization rules are changed.

The Content of Dictionaries Several classifications for information which is typically stored in relational distributed database dictionaries are [Bourne et al. 1986]:

1. Federation description. It includes the name of the global relations and sets of attributes.
2. Fragmentation description. In horizontal fragmentation (horizontal fragmentation consists of partitioning the tuples (records) of a global relation into subsets). In vertical fragmentation, it includes the attributes which belong to each fragment (the vertical fragmentation of a global relation is the subdivision of its attributes into groups.)
3. Allocation description. It gives the mapping between fragments and physical storages.
4. Mapping global names into a set of local names. It is used for binding the global names to the names of local data stored at each site.
5. Access control. It describes the access methods which are locally available at each site. For instance, in the case of a relational system, it includes the number and types of indexes available.
6. Consistency information. It includes information about the users' authorization to access the database, and/or integrity constraints on the allowed values of data.

Distribution of Dictionaries and Directories Dictionaries and directories can be allocated in the distributed database according to any of five schemes:

1. *Centralized.* The complete dictionary (or directory) is stored at one site; this solution has obvious limitations, such as the loss of locality of applications which are not at the central site; the loss of availability of the system, which depends on the single central site; and, finally, performance—the central site can become the bottleneck of the system.
2. *Fully replicated.* The complete dictionary (or directory) is replicated at each site. This alternative makes the read-only use of the

dictionary local to each site, but increases the complexity of modifying dictionaries, since it requires updating dictionaries at all sites.

3. *Local partitioning of dictionaries* (or directories). Dictionaries are fragmented and allocated in such a way that they are stored at the same site as the objects to which they refer. If there is a large number of non-local queries this alternative increases the volume of traffic on the network, and impedes the performance of the local system.
4. *Extended centralized dictionaries* (or directories). Every application program which requires a given part of the dictionary brings to its site the required information and keeps it during its execution; thus future references to the dictionary are likely to be local. The copies of the dictionary information are kept up-to-date by propagating updates to them.

Note. Chu [1976] has developed a mathematical model for comparing these alternatives. He reports that the ratio between update and queries to the dictionary turns out to be the most significant parameter. If the ratio is below 10 percent, then the preferred alternative is a distributed dictionary. If the ratio is between 10 percent and 50 percent, then the preferred approach is an extended centralized dictionary. Finally if the ratio is over 50 percent the preferred solution is the local dictionary.

5. *Caching the dictionary*. A final alternative, described briefly by Ceri and Pelagatti [1984], is practical approach: dictionary information which is not stored locally is periodically cached. This solution differs total replication in that the cached information is *not* kept up to date. In this case, a version number qualifies the cached dictionary information. If an application has tried to use a different version of the dictionary, then the up to date version is transmitted to the site, where the application tries to execute the data.

Site Autonomy The allocation and management of dictionaries is an important factor in determining the degree of local autonomy of each site. Indeed, one of the definitive features of site autonomy is that each site be capable of managing its own data, regardless of the other sites. Not all DBMSs have implemented the means of ensuring site-autonomy (INGRES·STAR and SDD-1, for example).

On the other hand, System R* [Lindsay et al. 1984] is a distributed database with full autonomy of sites. As a result, R* satisfies the following dictionary requirements:

1. Global replication of a dictionary is unacceptable, since it would violate the possibility of autonomous data definition.
2. No site should be required to maintain dictionary information of objects which are not stored or created on the site.
3. The name resolution should not require a random search of dictionaries entries in the network.
4. Migration of objects should be supported without requiring any changes in application programs.

R* meets these requirements by storing dictionary entries of each object in the following way:

1. One entry is stored at the birth site of the object, until the object is destroyed. If the object is still stored at its birth site, the dictionary contains all the information; otherwise, it indicates the sites at which there are copies of the object.
2. One dictionary entry, with a reference to the birth-site, is stored at every site where there is a copy of the object.

Thus, local references to data or other objects can be locally performed. On the other hand, if a remote reference is required, the systemwide name (the unique name given to each object in system R*), is used. If in the meantime the object has migrated to a different site, this information is returned to the requester, and one additional remote dictionary request will be required.

4.3 NETWORK DIRECTORIES

An important role for the network directory in an open system is to insulate the user and the user's application program from the characteristics of the communications network, and to allow users and applications to be moved among processors without impacting other users and applications that communicate with them.

Two trends point to the current developmental state of manually maintained directories: as networks become larger, the frequency of directory updating increases and the number of directories that must be consistently maintained grows. Furthermore, the trend toward peer connectivity among small systems requires that small systems also maintain directories—where in the past they may have depended upon a large host directory. The results are many more directories and increased update activity in a network. There is an obvious and growing need for the automatic and dynamic maintenance of directories.

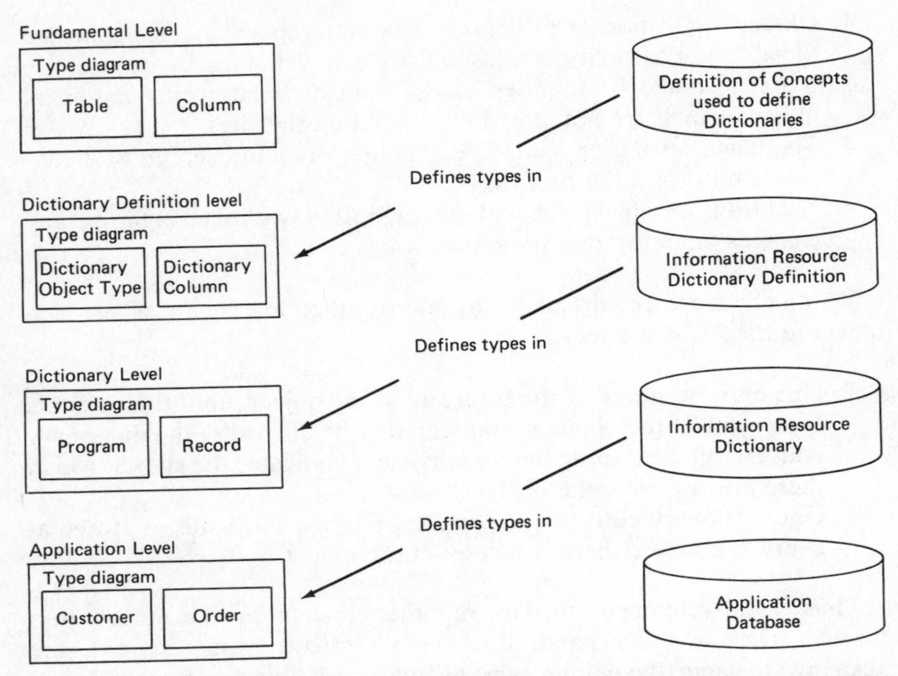

Figure 4.1. ISO levels definition of the Information Resource Dictionary System (IRDS). (*Source*: Bourne, "ISO/TC97/SC21/WG3, Information Resource Dictionary System, Part 1—Core," June 1986)

4.3.1 OSI Directory/Dictionary, Reference Model

The proposed OSI models [Bourne et al. 1986, ANSI 1985, ECMA 1984] for directory and dictionary systems is intended as a tool to assist in the development of systems and services. The suggested standards define several different functional components which work together. Applicable to a number of different physical and logical configurations, the models presuppose the use of the OSI Reference Model for open systems interconnection, and therefore the existence of a formally defined communications structure linking the components (see Chapter 6).

An important aspect of the ISO dictionary model is its formalization of the concept of *level of data,* as shown in Figure 4.1. Although the levels are interrelated, they exist to serve various different purposes:

1. Data on the *application* level is for the direct use of the information system's users.
2. The *dictionary* level contains data about the information resources.
3. The *dictionary definition* level contains data about the facilities

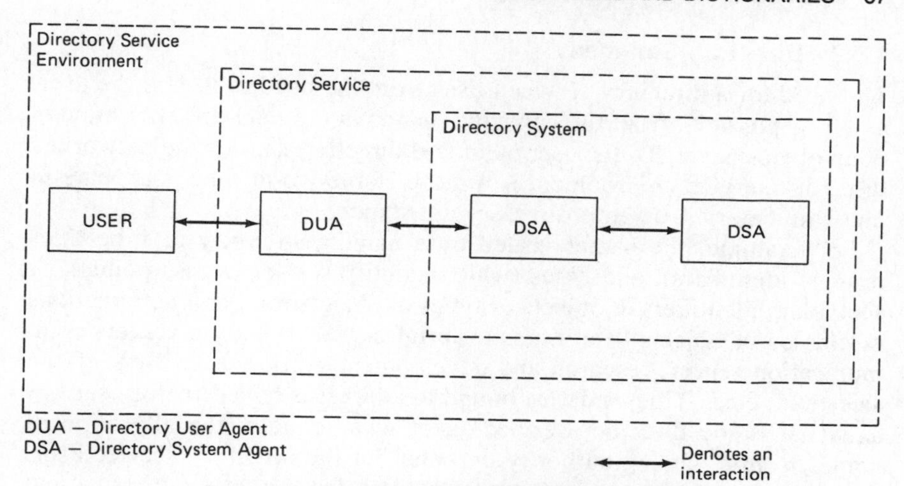

Figure 4.2. OSI functional model. (*Source*: ECMA)

provided by the Information Resource Directory System (IRDS) at the dictionary level. This level defines the types of data which can be maintained on the dictionary level, and also the consistency rules such data must obey.
4. The *fundamental* level exists solely to permit definition of standard facilities on the dictionary definition level, and to permit extension of these facilities by the introduction of subsequent standards.

Figure 4.2 presents a functional view of the OSI model for a *directory system*. This system consists of a number of directory system agents (DSAs), which communicate as necessary with one another in order to provide services to the clients of the directory system, and of directory user agents (DUAs), which reside in the local system of each client. The DUAs handle the protocol for communication between a user and a DSA.

A centralized directory system is a special case in the OSI model: there is only one Directory Service Agent. It seems unlikely that a centralized implementation would be able to fulfill the functional and organizational requirements of an OSI environment of any complexity. Directory systems lend themselves to distributed implementation and decentralized control, in which:

1. The primary directory information for different names can reside at different sites.
2. Locating a site containing the primary directory information for a given name can involve other sites, and the use of the name (or part of it) as an index to information residing at those sites.

4.3.2 Directory Services

The need for a directory service arises from the need to isolate the user, as far as possible, from the constant change in the network environment of an open system. To the user of an OSI directory service the networked directory service environment is a more stable world than it appears to one who enters it without directory assistance.

For example, the names issued by a naming authority may be titles (unique identifiers), addresses (which identify service access points), or secondary identifiers of objects or types of objects (i.e., characterizations of classes of objects—for example, print server, database server, communication server, research and development group within Geac, Geac salesman, etc.). Titles must be bound to addresses by a directory service agent. It is the directory service agent who requests the allocation of names from a naming authority on behalf of the directory service user. The directory service agent requests titles from a title authority and addresses to which the titles are to be bound from an addressing authority. The title authority and addressing authority may or may not be the same organization.

When the physical location of an object in the network changes, the user of that resource is isolated from the effects of those changes provided that a "name" rather than a physical address has been used to identify the object.

Another aim of the OSI directory services model is to provide a more "user friendly" view of OSI networks. For this reason directory services based on this model permit the user to obtain a variety of information about resources. The model supports:

1. The use of primitive forms of titles
2. The use of descriptive forms of titles
3. Synonyms for titles
4. Generic titles
5. Synonyms for addresses
6. Generic addresses
7. Multicast addresses
8. The binding of titles to addresses, which may be:
   (a) one-to-one
   (b) one-to-many
   (c) many-to-many
9. Operations for interrogation of the directory service in a user-friendly way to obtain information on primitive and descriptive titles
10. Operations for interrogating the directory service using primitive or descriptive titles as input, in order to obtain the addressing infor-

mation bound to those titles, and any information associated with the addressing information such as application-context data associated with an application-entity

4.4 NAMING

Naming systems are at the heart of open system directory system design. In spite of the importance of naming, no general unified treatment of them exists, either across computer science or within sub-areas such as networking, programming, operating system, directories, or database management systems. The central point in this chapter that we wish to make is that naming is not a topic to be treated lightly, in an ad hoc manner or in isolation from other open system needs.

Names are used for a wide variety of purpose within the open system such as: referencing, locating, scheduling, allocating, error controlling, synchronizing and sharing of objects or resources (information units, communication channels, transactions, processes, files, etc.).

Naming exists in different forms at all levels of open system architecture. Shoch [1978] has made a useful informal distinction between three important components of names widely used in distributed systems: names, addresses, and routes:

> "The name of a resource indicates *what* we seek, an address indicates *where* it is, and a route tells us *how to get there*."

A *name* is a symbol, usually a human-readable string, identifying some object or set of objects. The string need not be meaningful to all users and need not be drawn from a uniform name space. The string can identify processes, places, people, database, files, attributes, machines, and functions. Note that a name need not be bound to the address until this mapping takes place. The address (or addresses) associated with a particular name change over time.

An *address* is the data structure whose format can be recognized by elements in the address domain, and which defines a fundamental addressable object. An address in this context must, therefore, be meaningful throughout the domain, and must be drawn from some uniform address space. This address space may be a "flat" one which spans the entire domain, such as social security number, or it may be a hierarchical address space such as telephone or zip code numbers.

At the time one wishes to communicate with a particular address, there will be some algorithm that will map an address into an appropriate route. Note that the address need not be bound to the route until this mapping takes place; the choice of an "appropriate" route may change over time.

A *route* is the specific information needed to forward a piece of information to its specified destination. The routing action may only require one step to reach the specified address destination (a direct route) or it may require a series of steps in order to forward the information on its way. When the path to an address requires several steps (as in a store and forward system) the route defines a path through intermediate switching point.

Two levels of naming are required in an open system directory, one machine-oriented and the other intelligible to the human users.

4.4.1 Unique Machine-Oriented Naming

It must be possible at some level of the system directory to uniquely identify every object in the open system. The question is how to create a global unique name space in a heterogeneous environment, where each local identification domain may use one or more different schemes to create names unique within itself. Wheras global names are meaningful within a global, open system context, local names are meaningful within local contexts. One strategy for creating a global name space, called "hierarchical concatenation" by Pouzin and Zimmerman [1978], is to create a unique name for each identification domain, usually a server or host computer, and then to map it to the unique name for the domain. The other strategy is to develop a standard, uniform global name form and space for all resources, and to partition this global space among the local domains, which then map their local names to these global names. Binding of local names to global names can be permanent or temporary. This mechanism introduces a level of indirection with the advantages desired, e.g., local names can be chosen as most appropriate for local system directory, several local names can be mapped to the same global system directory or vice versa, and the global name can be reassigned if objects are relocated.

4.4.2 Human-Oriented Naming

Higher-level naming contexts are required to meet the needs of human users for their own local mnemonic names, and to assist them in organizing, relating, and sharing objects. Higher-level naming conventions and context can be embedded explicitly within the system directory. Mapping a high-level name to a machine-oriented identifier can be through the explicit context of the machine-name or implicitly by algorithms using the structure of the name. An example of a higher-level naming mechanism embedded in system directory is the mechanism used in the ARPANET mail service [Mochapetris 1983].

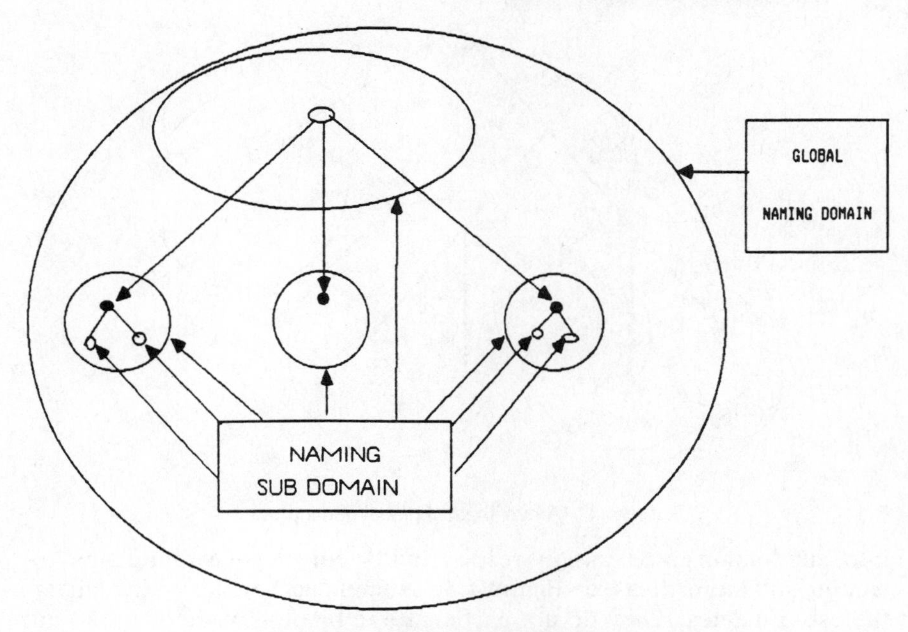

Figure 4.3. Naming domain structure.

4.4.3 Name Domain Structure

A name domain is any subtree of a global naming tree. Although one name domain may contain another, no two name domains may overlap. Figure 4.3 illustrates the OSI name domain structure [ECMA 1984]. Note that the fundamental structure of almost all name domains is hierarchical.

The advantages of hierarchical structure derive from the fact that the name space is partitioned. The size of individual directories is reduced and each directory may be maintained by a different server—perhaps on a different host. Partitioning also provides the means for easily grouping names relative to particular users. On the other hand, such partitioning can result in lower performance than using a flat name space. Consequently, some systems restrict the depth of the hierarchy (e.g., Clearinghouse).

4.4.4 Naming Domains and Authorities

A naming domain is the set of all possible names that may be assigned to a particular type of object. A naming authority is established to register unique names taken from a naming domain to be applied to a specific type of object. A naming authority may partition the naming domain

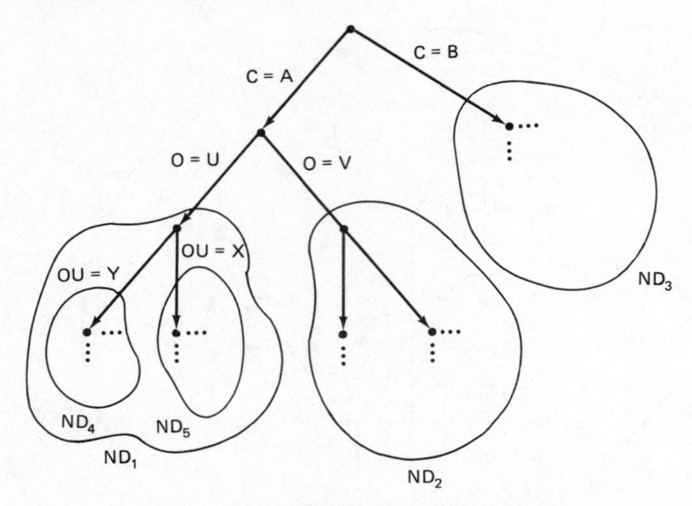

Figure 4.4. A hypothetical global naming tree.

into subdomains and assign responsibility for these subdomains to naming subauthorities (see Figure 4.4). Naming authorities only register the use of names. They do not participate in binding of the name to an object.

The global naming domain represents the set of all possible names for this type of object in the open system interconnect environment (OSIE). Independent global naming domains may exist within the OSIE for other types of objects.

Names taken from different domains either within or outside the OSIE may be bound to the same object. It is in this way that synonyms arise. Within a hierarchy of naming authorities the operation of each authority is independent of that of the other naming authorities at the same level, subject only to any common rules imposed by the parent authority.

A user of a naming authority may request an allocation of names from a naming authority, leaving the choice of names to the naming authority. Alternatively, a user of a naming authority may request an allocation of particular names. The naming authority may grant the request if it chooses. The user may interpret the names issued by a naming authority in any way it chooses.

4.5 IMPLEMENTATION ISSUES

It is important that the following concerns be addressed in the implementation of a directory for any open system environment. There are others specific to particular types of directory, but these are universal.

1. *Efficient support of transactions* (as well as session oriented services and applications), which implies minimizing, at all levels of the directory, the delay and number of messages that must be exchanged before and after the actual message is sent to perform the desired function. With respect to a system directory within open system, this means that we want to minimize the number of messages that must be exchanged to map a directory object held in a local space directly into a directory object usable at the global level.
2. *Support for at least two levels of naming,* one convenient for people and the other for machines. Properties desired in these two forms of naming are somewhat different. A machine-oriented name should be a bit pattern efficiently manipulated and sorted by machines and be directly useful with protection. A human-oriented name, on the other hand, is generally a human readable character string with mnemonic value. At some point the system directory must be used to map the human-oriented names into machine oriented ones.
3. *The generation of unique name structures* in a distributed manner. Directory systems are required to support unique names, where the structure itself need not be visible to the user.
4. *A naming mechanism in the directory* independent of the physical connectivity or topology of the system.
5. *A mechanism for the relocation of objects*. The implication here is that there should be at least two levels of naming, a name and an address, and that the binding between them should be dynamic. Also implied are algorithms for updating the appropriate directory context for the required mapping when an object is moved.
6. *Support for multiple, logically equivalent generic servers*. The implication in this case is that a single directory item can be dynamically bound to more than one address. When mapping takes place, one object is chosen according to some set of criteria associated with network management or other constraints.
7. *Support for the use of multiple copies of the same object*. Multiple copies are expected to be required frequently in distributed system to achieve performance, reliability or robustness goals. If the values of the directory entry are to be read only, more relaxed constraints can be imposed. On the other hand, if the values are to be modified, tougher constraints must be imposed to maintain consistency among the multiple copies. (See Chapter 7.)
8. *Support for multiple local user-defined names* for the same directory entry. This requirement requires that there be unique global

names for directory entries, and at least one level of local naming. Algorithms to bind local and global names are also required.

9. *Sharing of a single name by several different objects.* To support broadcast, or multicast naming for conferencing, at least two levels of identifiers and one-to-many mapping are required.
10. *The sharing of two or more resources, without naming conflicts.* This requirement implies a need for local identification domains for each object and context-switching mechanisms to bind local and global identifiers.
11. *A means of dealing with the reluctance of users to employ long absolute names.* An enrichment facility should map relative names into absolute names. (Sollins [1985] discusses this issue in detail.) Enrichment facilities can be thought of as ways to allow user to use incompletely specified names. In some situations, the user may remember only partial directory information and therefore require a "wild-carding" facility. Wild-carding support can reduce the amount of interaction between client and name service required to obtain a complete response to a query, but it also shifts much of the computational burden to the name service.
12. *There are at least two ways of implementing directory services, "integrated" and "segregated."* In integrated directory services, names for an object should be managed by the same object manager that implements the object. Integration is an example of the "end-to-end" argument. Let each "application" provide the minimal services it needed, rather than impose a separate general-purpose facility. Maintaining consistency between the object and its directory entry is trivial, since both are managed by the same directory server. Further, an object is accessible whenever its object manager is; this might not be the case if objects were named through a separate name server and the name server were inaccessible.

    On the other hand, segregated services explicitly separate the name management facilities from the servers that actually implement the object being named, leading to the notion of "name server." In the integrated approach, each server and client must duplicate certain functions. Each server must have its own name parsing code. A segregated approach eliminates this redundancy. Note that segregation does not imply a centralized implementation. For example, ARPA, Clearinghouse, and System R* are implemented in a distributed fashion.

In addition, Lantz et al. [1985] identify the following capabilities which should be provided in a more flexible directory service:

1. A well-designed service will decouple its services (i.e., the way in which they are presented to the user) from any specific implementation.
2. Though the use of common protocols should be encouraged wherever possible, general mechanisms for dealing with heterogeneous naming services should be provided. This requirement implies the provision of a global protocol translator.
3. The directory should have the ability to manipulate large classes of objects using common object manipulation protocols. In addition it should be able dynamically to create objects of new types and to manipulate them to the extent that the object manipulation protocols allow, without major changes to the system—such as recompilation or relinking. In general, such additions should not require human intervention.
4. The directory should support flexible protocols for external names consisting of sets of attributes and values permitting the user to attempt to name objects in terms of any information he may have available, rather than relying on a specific syntax.
5. It should be possible to associate an arbitrary action with a name. The action would be invoked whenever the name is. This action is in addition to any other information that may be associated with an object component—such as its type or access controls. This functionality can be used in combination with object types to implement, for example, extended protection mechanisms, performance monitoring, and cross domain naming.
6. The failure of any site participating in the naming service must not prevent any other site from accessing information about objects not stored on the failed site. It must be possible for individual system administrators to decide the best implementation technique for their system, in particular, an integrated or segregated approach. It should be possible to add the naming service to existing operating system without modifying existing code.

4.6 THE STATE OF THE ART

Over the past few years a number of name services for dictionaries and directories have been implemented for use in distributed systems.

In early message-based systems rudimentary name servers were developed that mapped simple string names for services such as "file system" into the identifiers for the processes that implemented those services [Baskett et al. 1977, Lantz et al. 1985, Zwaenepoel and Lantz 1984]. Similarly, restrictive name servers include those that have been developed to map string names for hosts or mailboxes into their network addresses. [Birrell and Nelson 1984, Oppen and Yogen 1983, Solomon et al. 1982] and the dictionaries of many databsae systems [Allen et al. 1982, Lindsay 1981]. More recently, several efforts have extended the notion of file system directory to include access to objects other than files, typically by having the directory entry contain a process or port identifier rather than a file identifier [Leach et al. 1983, Rashid 1980]. Other efforts have been oriented toward providing access to objects based on the attributes of the objects rather than by a fixed format name [Craft 1983, White 1984].

In the remainder of this chapter we examine a number of naming systems that represent the latest work in this area.

4.6.1 R* Dictionary Manager

System R*, a distributed database management system, has been developed by IBM's San Jose Research Laboratory. The database *dictionary* manages information on database objects, including their structure, format, access paths and access controls [Lindsay 1981]. Names are managed by a distributed collection of dictionary managers. Dictionary information about an object is stored at the same site(s) as the object itself. If an object is moved from the site at which it was created, its birth site, a partial dictionary entry is maintained at the birth site indicating where the full dictionary entry can be found. The object can be accessed directly at its new site without reference to the birth site, so that access to an object is still possible as long as the site that stores it is operational.

A name under system R*, referred to as a system-wide name (SWN), contains four components:

1. The user ID of the object creator
2. The user site of the object creator
3. The creator-specified object name of the object
4. The object-site or "birth site"

Object-autonomous sites can supply the first three components as they see fit, but site names must be unique. An SWN maps to information such as level, (storage) format, access procedure, and object type.

4.6.2 Sesame and Spice

Sesame [Jones et al. 1983], developed by the Computer Science Research Laboratory at Carnegie-Mellon University, is the file system for the Spice distributed operating system [Ball et al. 1982]. As with several other systems, its name service evolved from UNIX. The name service consists of a distributed collection of central name servers residing on the file server machines and Spice name servers residing on each user's workstation. The names for objects that are intended to be shared should be kept in directories that are always maintained by a central Spice name server. Names for objects employed primarily by one user may be kept in directories maintained by the user's local Spice name server. The directory entry associated with user-defined type is of fixed length but uninterpreted. There is no support within the Spice name service for guiding applications in the interpretation of user-defined types.

4.6.3 Universal Directory Service

UDS was developed and implemented by Computer System Laboratory, Stanford University [Lantz et al. 1985]. It integrates the following unique features:

1. The design was focused on protocol rather than hard code implementation. The benefit is that keeping information on protocols allows greater type independence. The UDS keeps a list of servers providing translation into a protocol as part of the protocol's directory entry.
2. The UDS can register name bindings for arbitrary object types—together with information on how to access and manipulate the object via media access and object manipulation protocols, respectively.
3. The UDS can both distribute and replicate partitions of its directory.
4. Sites may retain autonomy, both in operating in isolation and in enforcing local policies to control local resources.
5. One of the original contributions of UDS in the area of name management is the concept of the portal, which enables the system, for example, to:
   (a) Incorporate existing name spaces, even those with different name syntax
   (b) Create powerful context facilities
   (c) Enforce specialized access control schemes whenever needed
   (d) Transparently interpose monitoring facilities
6. The UDS design is geared to identifying and providing the information needed to write applications which are type-independent.
7. The UDS implementation supports the mapping of attribute-oriented "flat" names onto its hierarchical name space by providing a wild-card facility to allow searches based on attribute values. Thus, servers can provide users with a naming interface that is not positional in nature.

4.6.4 SNA Implementation of the Network Directory

The IBM System Network Architecture was developed during theearly 1970s. SNA [Sundstrom et al. 1985] distinguishes between resource names and their addresses. A name is a relatively stable identifier that users and applications can apply to other users or application programs.

By contrast, an address can vary according to operations decisions made in the network. Users and their application programs access resources by name. The system uses the name as a key to a directory that provides the current address of the requested resource.

Resources such as application programs, files, and logical units (LU) need to be registered only at their local (home) directories. Automatic network-wide searches of the various directories eventually result in the finding of the resource if an active path to it is available. SNA permits the replication of directory entries throughout the network, with several attendant benefits:

1. Performance in finding resource has been improved.
2. Switched links can be activated in order to complete required synchronous connections.
3. Asynchronous distributions can be forwarded as far as possible into the network when an active path to a destination is not available.

To meet the requirement for more dynamically maintained, distributed directories in an SNA network, new protocols will be needed for resource registration and network searching. These protocols will apply between directory users and their directory providers, and among the directory providers themselves.

4.6.5 ARPA Domain Name Service

A new name service for the ARPA (Advanced Research Projects Agency) Internet has been specified and is currently being implemented by the Network Information Center of SRI International [Mockapetris 1983]. It will run on a widely heterogeneous collection of machines, running a variety of operating systems. The domain name service is intended to help clients locate servers for common network services and to locate objects managed by such servers. The name space is hierarchical with no limitation on the depth of the hierarchy. The syntax is uniform across the entire name spaces. Each component of the hierarchy is a domain and typically reflects administrative or geographical grouping. The design allows an administrative entity to control what names are introduced into domains under its control.

Associated with each label in a domain is a set of *resource entries* that contain information about objects within the domain. Each entry contains a domain name and a number of other fields.

Name service functions are divided between two classes of "servers," name servers and resolvers, and are required to contain and use some knowledge of type (i.e., resource type defines a standard abstract resource such as "host address" or "mail forwarder") for their proper operation. Name servers are expected to recognize that certain type codes represent supertypes of other types. For example, a name server is expected to know that a request for objects of type MAILA (Mail Agent) can be satisfied by object of either type MF (Mail Forwarder) or MS (Mail Server).

4.6.6 V-System

The V-System has been developed and implemented by the Distributed System Group of the Computer Science Department at Stanford University. It is a server-based distributed system that runs on a collection of hosts connected by a high speed local area network [Cheriton 1984]. In the V-system the name space is partitioned among servers. Each server is expected to implement the objects corresponding to the names it defines. This approach allows the servers implementing objects a great deal of autonomy. However, it could also produce a great deal of confusion for clients unless they thoroughly understand the syntax of the names and the semantics of the operations offered by a server.

4.6.7 Clearinghouse

The Clearinghouse was developed by the Xerox Office Products Division in the early 1980s. It is used [Oppen and Yogen 1983] primarily to name mailboxes, users, and servers (machines in the context of a local net-based distributed system). The name space of the Clearinghouse is managed by a collection of Clearinghouse servers. Names are organized into a three-level hierarchy of the form local name, domain name, and organization name. The Clearinghouse can store arbitrary information about any named object. If the proper information were stored, applications could in principle be handed a description of how to operate on an object of some newly created type. There are two principal drawbacks to the Clearinghouse approach:

1. It forces type knowledge upon the client, i.e., the agent getting the description of a named object must recognize which property name

refers to the object's type and must then be smart enough to figure out what to do with an object of that type.

2. The Clearinghouse doesn't care what information is stored. The information needed to find an object or its manager may not be present at all.

Chapter 5

Design and Implementation of Open Systems

As the logical architecture of an open system introduced in Chapter 3 illustrates, software plays a critical role in the realization of an open system. Hence, current and future approaches within software engineering are critical in the development, use, and evolution of Open Systems. The emergence of software engineering as an important technology has already had an impact on how we view software. For example, the notion of the software life-cycle has enabled us to understand the software development process much more thoroughly. The emphasis on software reliability and maintainability influenced the design of Ada. The realization of the importance of human-machine interaction has stimulated work on human interfaces and in artificial intelligence. Emerging trends in software technology have been touched upon in Chapter 2 and are explored further in Chapter 8. This chapter examines current approaches in software engineering relevant to open systems.

From a pragmatic viewpoint, the development of an open system presents many of the same challenges encountered in the development of any complex software system—concern with reliability, maintainability of software over a long timeframe. An open system, however, also brings unique problems: for example, it may embody several different computer systems, comprised of different hardware and different operating systems, may involve several independent databases, and it may be spread over a wide geographical area. In the specification and design of an open system one may have to deal with existing databases, foregoing the luxury of building from scratch more convenient specifications. Moreover, because an open system involves processes executing on several machines, the testing of an open system is difficult, and often requires the development of specialized tools.

The following section reviews the software life-cycle model, and identifies areas of special concern in the different phases of the life-cycle. Subsequent sections examine approaches to the design of distributed systems, with emphasis on human interfaces, languages, and tools for implementation, software integration, and debugging.

5.1 THE SOFTWARE LIFE-CYCLE

The software life-cycle is conventionally divided into six stages, as in Figure 5.1 [Boehm 1981, Ramamoorthy et al. 1984]. The aim of the first

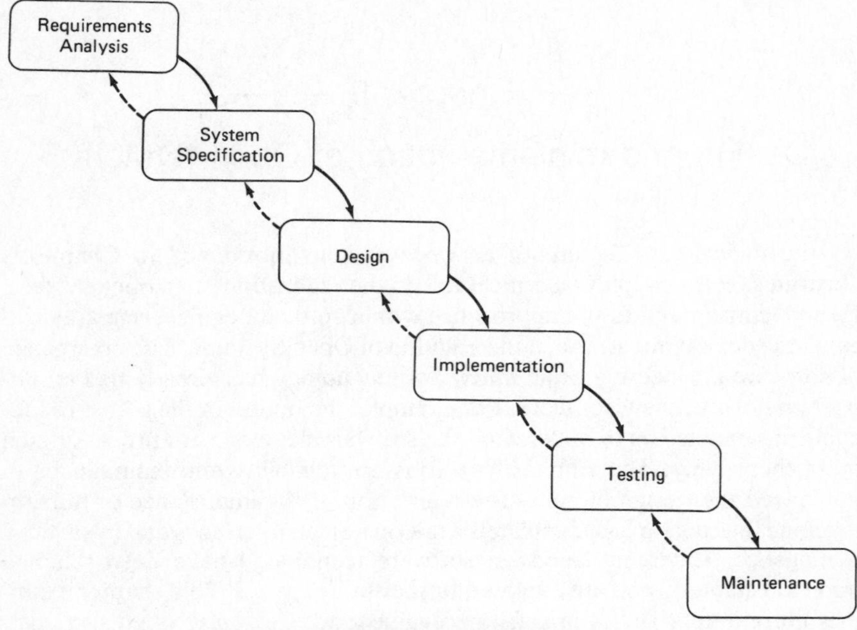

Figure 5.1. Stages of the software life-cycle.

stage is to understand the problem, that is to say, the requirements. Requirements are often derived from discussions with the "end-users"—those expected to use the final system. In the case of a distributed system, there may be additional requirements, not immediately related to the needs of users, which are imposed by existing software: existing operating systems, database management systems, utilities, and network standards. Hence, it is imperative to identify early in the requirements phase which existing software components will be retained and which will be rewritten.

The requirements, therefore, include the context in which the problem has arisen: constraints on the solution as well as the range of functions needed. In the light of these needs and constraints, management should be able to decide during this phase whether it is feasible and worthwhile to build the system.

A large percentage of the errors detected during testing, and a large portion of maintenance costs, are due to poorly identified and understood requirements [Boehm 1981]; the requirements phase is plainly critical to any project.

It is during the specification phase that software engineers attempt to define the external behavior of the system—what the system is to do, not necessarily how this is to be done. The specifications must be checked for consistency, omissions, ambiguity, and adherence to the requirements.

Because an open system may encompass a variety of hardware and operating systems, the specification of the system should be as independent of any particular machine and operating system as possible.

Although most specifications are written in "natural" language, it may be useful to specify some portions of a system either mathematically or by means of a specification language [Meyer 1985]. It may also be necessary to clarify requirements, and perhaps to iterate the definition of the specifications several times.

Once a stable specification has been formulated, the next task is to design the system. The purpose of the design phase is to specify in detail how the system is to be implemented. Since an open system must accommodate new hardware and software it is critical that the design clearly specify all interfaces, especially to such critical components as the communication system and databases; the development of internal standards would be useful at this point.

The design may be written in English, or expressed in a special design language. Tools to aid in the logical decomposition of functions may also be used. Because of the complexity of distributed systems—for example, the use of network communications, possibly heterogeneous databases, and concurrent processes—it may also be useful to model certain aspects at design time or to develop restricted prototypes of certain key components, for example, a distributed database. These aspects of the system, along with design methodologies and tools, are reviewed in Section 5.2.

Following design are the implementation and testing phases. Implementation should use a high-level, structured language, preferably one with constructs for implementing concurrent processes. Possible languages are surveyed in Section 5.3.

The purpose of testing is to find and correct any bugs in the system before it is released. Testing not only includes determining if programs behave properly, but that they are well-integrated, i.e., operate together. Testing distributed software presents a number of unique problems; some ways of dealing with these are presented in Section 5.4.

Once the system has passed testing it becomes operational, and enters the maintenance stage. A great portion of the maintenance costs arise from poor understanding of original requirements and faulty debugging.

5.2 SPECIFICATION

The specification of software systems is typically carried out in natural language, although more formal techniques are now coming into use. Interfaces with the human user must be considered early in the development cycle, and such interfaces are normally identified during the specification phase.

5.2.1 Formal Approaches for Specification

Formal approaches to the specification of software fall into three broad classes: algebraic specifications, denotational approaches and logical approaches. Algebraic specifications permit one to describe a system in terms of algebraic data-types, operations, and constraints [Guttag 1977, Guttag 1980]. The denotational approach involves specifying the software systems in terms of object domains and mappings or functions between the domains [Stoy 1977, Meyer 1985]. Such formal approaches are also related to work on automatic programming (see Chapter 2).

In practice on small software projects, these approaches have been proven to offer precision in defining specifications without loss of generality and conciseness. They have yet to be evaluated on large projects.

Several specification languages have recently been developed to support these approaches—e.g., SPECIAL [Roubine 1977] and SLAN-4 [Beichter et al. 1984]. The primary goal of such languages is to provide a syntax for expressing specifications, together with semantics for checking the consistency of specifications.

The third approach, the logical approach, makes use of a logic programming language, such as PROLOG (see also Chapter 2), in which specifications can be expressed [Kowalski 1979a, b]. Such specifications can then be "executed" to test for consistency. This approach may also provide a way to prototype software systems before design begins. This work also illustrates one area in which research within software engineering and artificial intelligence are interrelated.

All three approaches discussed have been applied to a variety of small software projects. Their use to specify completely a distributed system is questionable at this time, but they may be useful in specifying certain critical components.

5.2.2 Man-Machine Interfaces

It is now generally acknowledged that the man-machine interface of any software system is critical to the acceptance of the system by the end-user. The development of a distributed system is expensive and time-consuming; it is very worthwhile over the lifetime of the system to develop adequate man-machine interfaces for the users of the system. Ongoing work in artificial intelligence will certainly be of increasing relevance to interfaces (see Chapter 2).

Identification of User Models Before specifying any interfaces, the software engineers must identify the users of the end system. In general this will be very dependent on the use of the distributed system, its use in the corporate structure, the application software which depends on the distributed system, etc. However, it is possible to identify a number of broad classes of users: the application user, the application programmer, and the systems support personnel.

The application users will be those individuals interested in using the applications available on the system. This group of users would include secretaries, managers, accountants, production line managers, etc.

Application programmers and system support personnel are technical people responsible for, respectively, existing and new application programs and the maintenance and support of the distributed system itself. The database administrator or network administrator would both be considered as systems support personnel.

In each case it is important during the specification phase to identify these groups of individuals and the types of users within each group. The goal should be the development of a "user model"—a description of the user's background and assumptions about the user.

The identification of such groups is also important for subsequent design and implementation: the application programmers must know who will be using the applications and the system personnel need to know the requirements of the application programmers.

Guidelines for Man-Machine Interfaces Although the field of human factors in interactive systems is relatively young and no definitive rules exist, research in the past few years has enabled a number of guidelines to be identified. The following is a list of generally accepted guidelines [Wasserman 1982, Norman 1983, Shneiderman 1979]:

1. Make invisible the underlying aspects of the computer not necessary for the user to understand the software.

For users of applications, this implies that they see only what is relevant to their particular application. In the case of application programmers, this implies that they should only see those components of the system necessary to develop and maintain applications. This may imply that they have limited access to databases and network facilities. This may also be very important in maintaining the security and integrity of the distributed system. Finally, the systems personnel are, presumably, those individuals who developed the system and will be responsible for maintaining and enhancing the distributed system.

2. Make it virtually impossible to the user to cause the program to terminate abnormally.

This should always be the goal of any application software, i.e. it should be impossible for application users to cause any abnormal termination. This also implies that application programmers must keep this in mind during the development of application software.

3. Notify the user before any major consequences.

This implies, for example, that before a user or programmer deletes data or program, the system requests confirmation.

4. Provide on-line assistance.

Any interactive system must provide access to on-line information. Preferably, the user, especially the application user, should be able to request help at any point during the use of an application.
This would also suggest that appropriate facilities and tools be available for application programmers to support the development of on-line assistance in a distributed environment.

5. Tailor input requirements to user skills.

Clearly, certain users are capable of handling and perhaps prefer certain types of input. Casual users of applications may prefer menu selection and form entry; application programmers may prefer sophisticated screen editors. For these reasons, it is often advisable to discuss possible forms of input and interfaces with user groups during the specification phase. This can result in the identification of preferences and more detail in the user's models.

6. Minimize the work to be done in the event of a user error.

One of the most frustrating events in the use of any software system is making an error in entering a relatively lengthy input string and having the system request re-entry. When possible, the software should prompt for corrections or allow the user to easily change the string once entered.

7. Maintain a consistent response time.

Application users are easily frustrated by slow response times, especially for straightforward requests or when there are great variations in the time for similar or identical requests. It is important to identify acceptable limits on response time before system design begins.

These guidelines are far from complete, but serve to indicate the kinds of considerations that must take place in the development of appropriate interfaces.

Tools for Specifying Interfaces In recognition of the importance and difficulty of specifying and designing appropriate user interfaces, recent work has focused on tools to aid in these phases.

One approach, the formal language approach [Reisner 1981], requires that the interface be specified by means of a Backus-Naur Form (BNF) grammar. This allows the designer to specify unambiguously the structure of an interactive dialogue. An alternative approach, using transition diagrams, requires the designer to specify the interface in terms of states and transitions [Wasserman 1982; Jacob 1981, 1983]. In both approaches machine processable descriptions are possible, thus allowing tools for analyzing interactive dialogues and for generating prototype interfaces. Such approaches tend to be more applicable to certain application areas than others, e.g., data entry and retrieval rather than text editors.

5.3 DESIGN OF DISTRIBUTED SYSTEMS

Once specification has been completed, project development moves into the design phase. The goal here is to decompose the specifications into procedures, modules, data structures, interfaces, etc. based upon some criteria, such as functionality, data flow, portability, etc. It is also during this stage that models or prototypes of components of the system being designed may be developed.

5.3.1 Design Methodologies

Current methodologies for designing complex systems can be grouped into three broad categories. The first may be referred to as *functional decomposition*. This involves a step-by-step division of the functions within the specifications into subfunctions, etc. It provides a very good way to develop a hierarchical design and is most appropriate for applications which are well-understood. Hence, it may be most suitable for the design of well-understood components of a distributed system, e.g., known applications, editors, compilers.

The *data-flow design* approach decomposes the system into modules. For each module one specifies the data which the module processes and produces. The modules are generally classified as input, output, or transformational, depending on whether the module receives input from external sources, produces output for external targets, or processes the data. Note that input or output could be to or from another application.

This design approach is particularly useful in the design of distributed systems because of the emphasis on data flow; the nature of distributed systems require that they involve a number of asynchronous processes communicating via messages. It may, however, be difficult or artificial to partition all functions into input, output, and transformation modules.

A third approach is the *data structure design* method [Jackson 1975]. This approach is particularly suited to those systems with well-understood data structures and produces programs based upon the hierarchical nature of these structures.

Of these, the data flow approach is most appropriate for designing distributed systems, especially for what one would call the "systems" layer. However, it should be remembered that such systems are complex and typically involve a number of applications, both for end users and for application programmers, and hence the other approaches, particularly functional decomposition, may be appropriate.

5.3.2 Models and Tools

As the complexity of software systems increases, software engineers are looking more to formal methods and tools to aid in the design process. The purpose of using tools and models at this stage of the development is to both document and evaluate various aspects of the design before implementation. Both can be used to determine consistency of designs, evaluate critical algorithms, obtain performance estimates, etc.

Petri Nets Petri Nets [Peterson 1977, 1981; Yau and Caglayan 1983] have been used to describe concurrent process for a number of years. Informally, a Petri net is a graph consisting of two kinds of nodes: *places* and *transitions*. Places are usually represented by circles and transitions by bars. Places can hold *tokens* which are represented by dots (see Figure 5.2). A *marking* of a Petri net consists of a number of markers placed at each place (possibly zero); the initial marking determines how the tokens are placed initially. Rules define how one marking leads to a new marking: a transition *fires* if all places with arcs to the transition contain tokens and then the transition removes the tokens from the input places and adds tokens to its output places.

Petri nets can be used to model both the static and dynamic properties of distributed or concurrent systems. Static properties can be determined by examining the graphical part of the Petri net. Dynamic properties can be determined with the use of the transition rules.

Petri nets offer a number of advantages for modelling distributed systems. They can be used to produce a precise representation of concurrent components. Analysis tools can be developed and used to

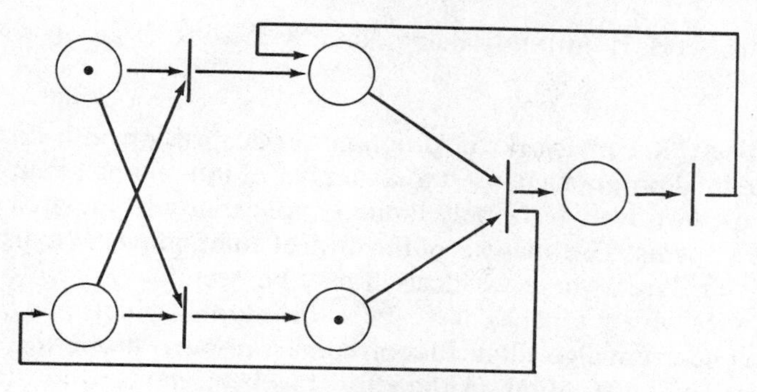

Figure 5.2. Example of a Petri net.

examine and verify dynamic properties of the systems. The nature of Petri nets makes them most useful for modelling concurrent, nondeterministic or asynchronous events. A number of extensions to Petri nets have been proposed [Nutt 1972, Noe and Nutt 1973, Noe 1978, Yau and Caglayan 1983] for modelling additional aspects of distributed systems.

Language-Based Approaches A language-based approach is an alternative to the graphical one offered by Petri nets. Like the development of design languages for more common software systems (e.g., PDL), [Caine and Gordon 1975] there have been a number of efforts to develop languages and systems for designing distributed or concurrent systems.

DREAM (Design Realization, Evaluation, and Modelling) [Riddle 1978, 1980] is a project oriented toward the design of concurrent systems. The heart of DREAM is the DDN (DREAM Design Notation) language. It provides a method for describing systems hierarchically and modelling such systems as collections of components communicating directly via message transmission or indirectly via information areas.

The complete system also includes a database for storing textual descriptions, bookkeeping tools, editor and syntax checker, information extraction and report generation tools, and tools for checking for various incompatibilities between components.

SARA [Campos and Estrin 1978, 1982] is a design methodology which uses both language descriptions and graphical representations to describe systems. COSY [Lauer 1979] is a design specification language based upon path expressions.

Ada [ADA79, ADA80] as a programming language offers constructs for abstraction (see Section 5.4.2), modularity, and concurrency (see Section 5.4.3). Such constructs are useful within a design language and are being explored as the constituents for an Ada-based design language [IEEE

1984]. This work is currently being done by a working group within the IEEE.

Verification Recent work on program verification of both concurrent and sequential programs has led to a number of interesting results. These techniques, however, have only limited applicability in the area of large software systems. The analysis of the critical components of a distributed system is one area where verification may be useful.

A number of approaches have been developed which allow one to analyze concurrent algorithms for correctness or other properties, such as deadlock avoidance. Much of the effort has been devoted to the formal analysis of concurrency control algorithms—particularly important in maintaining the integrity of distributed databases.

Verification of concurrent algorithms has been successfully accomplished using several different approaches. Graphical approaches [Bhargava and Hua 1983, Chen and Yeh 1983, Voss 1980], often relying on Petri net representations, can be used to establish the correctness of algorithms as well as other properties. Approaches based upon logic [Pnueli 1977; Owicki 1976, 1982] are also very general, but require the specification of program invariants which are often difficult to specify and understand. A third approach is based upon behavior traces [Hoare 1978; Misra and Chandy 1981] of communicating processes. This approach assumes a total ordering of events within a distributed and certain properties of concurrent algorithms cannot be verified [Misra and Chandy 1981]. Other approaches based on formal grammars [Ellis 1977] and event paths [Campbell and Haberman 1974, Lauer and Campbell 1975] have also been developed and used verify properties of concurrent algorithms.

It is currently impractical to attempt to verify the correctness of the design of a distributed system. It may be valuable, however, to attempt to verify certain properties of critical algorithms, e.g., concurrency control, of the system at design time. This can be a useful exercise in uncovering unsuspected problems.

5.4 LANGUAGES AND IMPLEMENTATION OF DISTRIBUTED SYSTEMS

At a time when software costs are increasing and applications are growing more complex, the thought of implementing a distributed system or distributed applications can weaken the heartiest of programmers. Firstly, such systems are inherently large and complex, often consisting of many interrelated applications and subsystems. Secondly, the programmer must deal with new concepts, such as concurrency, and cope with extra constraints introduced by a distributed system, such as communication failures. Languages for implementing such systems and applications have been developed with the goal of easing the task of the programmer. This chapter reviews the basic concepts and several languages.

5.4.1 Basic Concepts

Two key concepts are fundamental in the successful design and implementation of distributed systems: abstraction and concurrency. The latter embodies the problems and solutions for coping with programs communicating with one another while executing on different machines. Abstraction refers to the task of identifying and specifying application entities and their set of allowable operations. The following expands on each of these basic concepts.

Concurrency In a broad sense, a distributed system can be considered to be a collection of independent processors communicating via some medium. A *process* can be considered to be a sequential program which executes on a single processor and which can communicate with other processes. Processes communicate via *messages*.

Two processes are said to *execute in parallel* or *execute concurrently* when the execution of one does not depend on data (including messages) from the other. This does not exclude two parallel processes from sending each other messages, i.e., it implies that each would execute regardless of whether the messages were sent or not. Alternatively, two processes whose execution is dependent upon each other are often called *cooperating* processes. Because of the dependencies, such processes must coordinate their execution (e.g., send a message and wait for an acknowledgement); such coordination is often called *interprocess synchronization* since the sequential execution of each process must be coordinated with the other. The code to be executed within each process during the period of synchronization is often called a *critical section* or *region*. During the execution of the critical section the process cannot receive any messages; hence such code is usually kept to a minimum.

It is the implementation of cooperating processes which introduces many of the problems during the development of distributed systems and applications. It is also unfortunate that many of the proposed and needed applications of distributed systems require cooperating processes. For example, the successful development of a distributed database requires such cooperating processes to ensure the integrity and security of the database.

The implementation of cooperating processes requires that the programmer be aware of a number of possible problems: deadlock, nondeterministic events, process termination. *Deadlock* occurs when two processes are waiting for each other to send messages to the other:

neither can proceed. Because messages are sent over communication lines, noise or other interruptions can take place, many of which cannot be anticipated by the programmer: *nondeterministic events*. It may also happen that one process unexpectedly terminates, say due to failure of the processor running that process; other processes must be able to continue (or at least stop gracefully).

In order to facilitate the cooperation between one or more processes, the communication is usually defined in terms of a well-defined "language" or *protocol*. A great deal of effort has been devoted to the development and analysis of protocols for communicating processes.

Abstraction Abstraction within a programming language is useful in the implementation of any software. Firstly, it enables the programmer to define data objects in the software in a natural and readable way in terms of the objects that occur in the problem domain. Secondly, with the use of subprograms, the programmer can define operations on such data objects and, once defined, use the data objects without worrying about details of their implementation.

A programming language can provide constructs for abstraction at varying levels. It might only provide for type declarations, i.e., provide for the definition of new types of objects within a program, which are then global throughout the program (for example, PASCAL). In such a language, the programmer has access to the object's implementation and can circumvent the use of predefined subprograms. This, of course, leads to subsequent problems with debugging and maintenance of the software.

Alternatively, the language (such as Modula II or Ada) can provide abstraction constructs which allow the type definition and operations to be grouped together, effectively hiding the details of object implementation from the programmer. Manipulation of an object can only be done via the operations provided. This is often referred to as *data hiding*.

This can be further enhanced if the programming language provides facilities for the definition of object types and their operations in modules or units separate from programs which use them, i.e., provide some separate compilation facilities.

In the design and implementation of open systems, one is very much interested in providing layers of software which are used by applications, which are independent of the details of other layers and which hide details from programmers. Abstraction facilities can provide the language mechanism for achieving such layers. This, in turn, can encourage modular design. Details of concurrency, network communications, database access, etc., can be hidden from programmers and the needed capabilities still provided. Such constructs further permit detailed changes to such functions without affecting the implementation of applications.

5.4.2 Language Issues

From the viewpoint of the implementer, the language for implementing a open system must provide the basic constructs for dealing with the above concepts. In turn, the realization of such constructs within a programming language can take on many forms and be implemented in many different ways. The designer and/or the user of a language for implementing distributed programs should be aware of a number relevant issues:

1. *Language capability*. Are the constructs provided within the language sufficiently general? Can the given protocols be implemented? Are there restrictions on process synchronization or concurrency imposed by the language.
2. *Efficiency*. Are the constructs efficiently implemented? Are there hidden costs associated with certain constructs (e.g., buffers, message queues)? Are the schemes used for passing messages efficient?
3. *Abstraction*. Are the constructs sufficiently abstract so that excessive details are hidden from the programmer? Can the programmer encapsulate those data structures, types, and operations involved in process communication so that other portions of the program or application are suitably insulated?

In addition, the language for implementing a distributed system should be a high-level language satisfying what are now generally accepted programming language "standards":

- Data definition constructs supporting a variety of data types and structures, data type checking, and basic data abstraction facilities (user type definitions)
- A collection of structured programming constructs, e.g. IF-THEN-ELSE, WHILE, CASE, etc.
- Constructs for the definition of program units such as procedures and functions
- A capability for separate compilation of programs or program units

Although not necessary for the implementation of a distributed system, such language facilities can aid in the implementation by providing layers via libraries or pre-compiled units which hide details from application programmers.

5.4.3 Language Approaches to Abstraction

Various forms of abstraction can be found in most programming languages, the most common being the subprogram. Some languages permit the definition of user-defined types, e.g., PASCAL and C, but do not provide mechanisms for limiting the accessibility of such objects. More recent languages, such as Ada and Modula II, provide abstraction mechanisms beyond global user-defined types and subprograms. The following sections describe the abstraction mechanisms of these two languages; their features can then form a basis for comparison of abstraction mechanisms in other languages.

The abstraction mechanism in Modula II and Ada is based on the concept of a *module* (called a *Module* in Modula II and a *Package* in Ada). A module is a collection of related subprograms operating on a set of data types. The subprograms and associated data structures may be accessed by other programs. Details about the representation of data objects and the implementation of the subprograms may be suppressed.

Abstraction in Ada A package in Ada is a program unit containing related computational components, such as data types, objects, and subprograms. Packages may be used to encapsulate an abstract data type and its associated operations (e.g., a stack or complex number) and hide implementation details from programmers. A package may also be used to collect a group of related components, as in a collection of mathematical functions.

Ada permits the software designer to separate the specification of a package (i.e., its definition) from its implementation. The specification describes the accessible parts of the package and how they may be used. The implementation part then defines the actual implementation details. The specification part may be compiled separately from the implementation part. This encourages designers to develop package specifications early during software development, without requiring implementation details to be specified and not requiring other members to become aware of implementation details.

The basic structure of an Ada package specification is

```
package NAME_OF_PACKAGE is
  —visible definitions
private
  —hidden definitions
end NAME_OF_PACKAGE;
```

Components defined in the visible portion of the package specification are accessible outside the package. Definitions within the private portion provide information for the compiler (e.g., for storage allocation) and can be seen, but not accessed, by the user of a package.

The basic structure of the implementation portion is defined by a *package body*:

```
package body NAME_OF_PACKAGE is
  —definitions of functions and procedures
end NAME_OF_PACKAGE;
```

The following example illustrates the specification of an Ada package for stacks of integers:

```
package STACKS is
  type STACK(SIZE: POSITIVE) is limited private;
  procedure INITIALIZE(S: in out STACK);
  function IS_EMPTY(S: STACK) return BOOLEAN;
  procedure PUSH(ITEM: in INTEGER; ON: in out STACK);
  procedure POP(FROM: in out STACK; ITEM out INTEGER);
private
  type LIST is array(INTEGER range <>) of INTEGER;
  type STACK(SIZE: POSITIVE) is
  record
  OBJECTS: LIST(1..SIZE);
  TOP: INTEGER range 0..SIZE;
  end record;
end STACKS;
```

Ada provides private types and limited private types. In either case only the information available in the visible portion is available to the user. For private types the operations of assignment and test for equality and inequality are available. These are not available for limited private types unless explicitly provided for by the package. In the above, a user would be prohibited from comparing two stacks for equality.

To define a stack of 100 integer entries, one could specify:

```
STACK100 : STACK(SIZE → 100);
```

As noted, the above package specification is for a stack of integers. An application requiring several different types of stacks, e.g., stacks of integers, characters, names, would require three separate package definitions. To remedy this problem Ada provides *generic* program units; this approach takes abstraction of such data types as stacks to their full limit. A generic program unit essentially defines a program template which may be written once and tailored at compile time.

The following turns the preceding package specification for stacks of integers into a definition for stacks of any object type:

```
generic
  type OBJECT is private;
package STACKS is
  type STACK(SIZE: POSITIVE) is limited private;
  procedure INITIALIZE(S: in out STACK);
```

```
    function IS_EMPTY(S: STACK) return BOOLEAN;
    procedure PUSH(ITEM: in INTEGER; ON: in out STACK);
    procedure POP(FROM: in out STACK; ITEM out INTEGER);

  private
    type LIST is array(INTEGER range <>) of OBJECT;
    type STACK(SIZE: POSITIVE) is
    record
    OBJECTS: LIST(1..SIZE);
    TOP: INTEGER range 0..SIZE;
    end record;
  end STACKS;
```

To use a generic procedure, the procedure must be instantiated. This requires the user to specify actual parameters for the generic parameters. The following would create a new type of stack, one whose objects are letters:

```
  type LETTERS is ('a'..'z');
  package LETTER_STACK is new STACK (OBJECT → LETTERS);
```

This, in turn, could be used to define stacks of letters:

```
  LSTACK100 : LETTER_STACK(SIZE → 100);
  LSTACK10  : LETTER_STACK(SIZE → 10);
```

Abstraction in Modula II A collection of subprograms operating on a set of data types and objects is called a module in Modula II. As with packages in Ada, modules are split into specification portions (DEFINITION MODULEs) and implementation portions (IMPLEMENTATION MODULEs). Each portion may be compiled separately. As with Ada, the specification portion contains the definition of data types, objects and subprograms which may be used by another program.

The following illustrates the Modula II specification for stacks of integers.

```
  DEFINITION MODULE Stacks;

    EXPORT QUALIFIED
      (* types *)
    Stack,
      (* procedures *)
    Empty, Pop, Push, Initialize;

    TYPE Stack;
    PROCEDURE Empty(S: Stack): BOOLEAN;
    PROCEDURE Pop(VAR S: Stack): INTEGER;
    PROCEDURE Push(VAR S: Stack; X: INTEGER);
    PROCEDURE Initialize(VAR S: Stack);

  END Stacks.
```

Notice that the specification of the module Stacks reveals nothing of the implementation details. That is, the hiding of the implementation details in Modula II is much more strictly enforced than in Ada.

The components defined in the module Stacks must be explicitly imported to be used within another module, for example:

```
MODULE TestStack;
  FROM Stacks IMPORT Stack, Empty, Pop, Push,Initialize;
  .
  .    (*remainder of program*)
  .
END TestStacks.
```

The details of the implementation are defined in a separate module as follows:

```
IMPLEMENTATION MODULE Stacks;
  CONSTANT STACKSIZE = 80;
  TYPE
  Objects = ARRAY (1..STACKSIZE) OF INTEGER;
  Stack = POINTER TO RECORD
    ITEM: Objects;
    TOP: CARDINAL
  END;
  .
  .    (*implementations of procedures*)
  .
END Stacks.
```

Modula II requires that hidden types be pointer variables, which is the reason for the definition of Stack above.

The definition module for Stacks would have to have been compiled and available before TestStack could be compiled and executed. Providing that the specification portion of Stacks does not change, a change in the implementation module of Stacks would only require that module to be compiled; neither the definition portion of Stacks nor module TestStack would have to be recompiled.

5.4.4 Language Approaches to Concurrency

This section examines the concurrent programming constructs found in several different programming languages: Ada [Wegner and Smolka 1983; Ledgard 1981; Ada 1979, 1980], Concurrent C [Tsujino et al. 1984] and CSP (Cooperating Sequential Processes) [Hoare 1978]. Details about other aspects of each programming language can be found elsewhere.

It should be noted that cooperating processes have been used within operating systems for many years. In such an environment the processes reside on a single processor and can communicate via *shared memory* (e.g., *semaphores*) common to both processes. Languages such as C, Concurrent Pascal [Brinch-Hansen 1975], Modula [Wirth 1977] and Modula II [Wirth 1983] provide only interprocess communications via shared variables. Because of the shared memory requirement, such languages are not appropriate for distributed systems [Williamson and Horowitz 1984] with existing constructs. In the case of languages such as Modula II or C, for example, extensions could be provided (via Modules in Modula II and via library functions and the preprocessor in C). Such extensions would have to be implemented using the communication primitives provided by the operating system.

Ada Processes in Ada [Wegner and Smolka 1983, Ledgard 1981] are called *tasks*. Ada permits both static and dynamic task creation. The former is useful when the number of tasks to execute concurrently is known at compile time; dynamic tasks are needed for applications in which patterns of concurrency cannot be predicted.

In the following example, the execution of the body of the procedure THREE_TASKS is done concurrently with tasks ONE and TWO. At the start of the execution of the procedure body, the tasks are initiated and then executed. The procedure THREE_TASKS cannot be executed until the tasks ONE and TWO have terminated.

```
procedure THREE_TASKS is
  task ONE;
  task TWO;
  task body ONE is separate;    --defined elsewhere
  task body TWO is separate;    --defined elsewhere
begin
  .     --start execution of procedure body and initiate
           tasks ONE and TWO for concurrent execution
  .
end;
```

Task initiation and termination in Ada is therefore tied to the nested block structure of the program.

A task body in Ada has the form:

```
task body TASK is
     --declarations
begin
     --statements in task
end;
```

Initiation of a task causes the body to begin execution, beginning with the declarations. Ada permits the specification, body description and initiation of a task to occur at textually separate points of a program.

The following example illustrates the use of dynamic tasks and lists:

```
procedure CREATE_CUSTOMERS is
  task type CUSTOMER;
  type CUSTOMER_PTR is access CUSTOMER;
  type LIST_RECORD;    --incomplete definition
  type LIST_RECORD_PTR is access LIST_RECORD;
  type LIST_RECORD is
  record
  ID: INTEGER;    --customer id.
  C: CUSTOMER_PTR;    --ptr. to cust.task
  NEXT: LIST_RECORD_PTR;    --next rec. ptr.
  end record;

  HEAD, NEW_EL: LIST_HRECORD_PTR := null;

  task body CUSTOMER is separate;

begin
  for I in 1..10 loop—create a list of 10 customers
    NEW__EL := new LIST_RECORD'(I, new CUSTOMER, HEAD);
    HEAD := NEW_EL;
  end loop;

end CREATE_CUSTOMERS;
```

In this example, ten CUSTOMER tasks are created; a new task is created dynamically each time through the loop and added to the list of customers. The procedure CREATE_CUSTOMERS terminates when all tasks have completed execution.

Communication between tasks in Ada is done by synchronization followed by message passing (tasks may also communicate by shared variables). Ada tasks require a *specification* which defines the information to be used by other tasks to communicate with it. The specification itself may be in a textually separate segment from that of the task body. Consider the following example:

```
task BULLETIN_BOARD is   --specification of task
  entry POST (M: in MESSAGE);
  entry REMOVE (M: out MESSAGE);
end BULLETIN_BOARD;

  ...

  task body A_TASK is

  ...

  MY_MESSAGE : MESSAGE;
```

```
    . . .
    BULLETIN_BOARD.POST(MY_MESSAGE);
    . . .
  end A_TASK;
```

In this case, the specification of the task BULLETIN_BOARD, contains information identifying two task entries: POST and REMOVE. In the task A_TASK the task BULLETIN_BOARD is sent a message corresponding to the entry POST and a parameter MY_MESSAGE. The entry call can be viewed as a procedure call, but requires synchronization and communication between the calling and called tasks.

To complete this example and illustrate the latter, consider the following portion of the task BULLETIN_BOARD:

```
  task body BULLETIN_BOARD is
    . . .
    accept POST(M: in MESSAGE) do
    . . .     --accept body for POST
    end;
  end BULLETIN_BOARD;
```

The execution of BULLETIN_BOARD.POST(MY_MESSAGE) causes the synchronization between the call and the accept statement within the task BULLETIN_BOARD. Once synchronized, communication takes place via the parameter: MY_MESSAGE is sent to task BULLETIN_BOARD. The body of the accept statement is then executed. Once this has been completed, the calling and called tasks may resume execution. Any parameters to be returned are returned after the execution of the accept body and before the two tasks are permitted to proceed independently. Note also that only the calling task, i.e., A_TASK, need know the name of the other task; BULLETIN_BOARD need not know the name of any caller. This is preferable to having both tasks know the name of each other.

Each task entry has an associated queue for waiting calls that have not been serviced. By ignoring parameters, this process can be used to synchronize two tasks.

Ada also provides the Select construct for the nondeterministic scheduling of concurrent processes. The Select construct is used in conjunction with conventional constructs for conditional branching and looping. The following illustrates the structure of the Select construct:

```
loop
  select
    when Guard_Condition_1 →
  accept ENTRY_H1 do
  . . .    --critical section 1
  end;
  . . .    --code for entry ENTRY_1
  or
    when Guard_Condition_2 →
  accept ENTRY_2 do
  . . .    --critical section 2
  end;
  . . .    --code for entry ENTRY_2
  or
    terminate;
  end select;
end loop;
```

In the evaluation of the Select statement, the guards (GUARD_CONDITION_1 and GUARD_CONDITION_2) are evaluated to determine which are true (there may be more than one). Those that are read to accept are then considered. If there is only one alternative which is true and ready to accept, then execute it. If more than one is true and ready to accept, then one of the alternatives is arbitrarily selected, i.e., is implementation dependent. Otherwise, wait for the first true alternative that becomes ready. A Select statement must have at least one branch that is an accept statement.

Ada requires an explicit terminate alternative (via the terminate statement). Let the terminate alternative in question be contained in a task T1 and let T1 be dependent upon some program unit P. Let T2, . . ., Tn also be tasks dependent upon P. Then the terminate alternatives is taken only if all tasks dependent on P, i.e., T1, . . ., Tn, have terminated or are waiting a terminate alternatives.

Other discussions on the Ada Select statement can be found in a number of papers [Wegner and Smolka 1983, Pnueli and DeRoever 1982, Francez and Yemini 1982].

5.4.4.1 Concurrent C

Concurrent C is an extension of C [Tsujino et al. 1984] which contains extensions to make it more suitable for the implementation of distributed systems. C, of course, is a systems programming language with capabilities for process creation and interprocess communication via pipes (process communication channels) and signals (software interrupts). Concurrent C extends C to handle parallel processing and interprocessor communication.

The basic constituents of Concurrent C are *program units* and *processes*. A program of Concurrent C consists of several program units and processes running on them.

The program unit of Concurrent C has the same structure as a C program and exists on a single processor. A C program is treated as a Concurrent C program with a single program unit and a process.

A process in Concurrent C is similar to a function definition within C. The following illustrates the definition of a process which implements a message buffer, where a message is a sequence of characters, between two other processes:

```
process buffer(sender, receiver)
  processid sender, receiver;
  {
    char msg[N];
    int in=0, out=0;
    forever
      select {
    case in != out+N
    receive(msg[in++%N]) from sender;
    break;
    case in != out
    receive() from receiver;
    send(msg[out++%N]) to receiver;
      }
  }
```

When the buffer process is *activated* (see below), the process identifiers of sender and receiver processes are passed to it. The outer set of brace brackets define the process function body which is executed when the process is activated. The process terminates when its execution reaches the end of its process function; in this case, this process loops forever or until killed by a parent process. Like functions within C, process functions may not be nested.

A process is *activated* by means of an active statement. The above process might be activated by means of:

```
activate Buff_Pid = buffer(Pid1,Pid2);
```

where Pid1 and Pid2 contain the process identifiers (represented by integer values) of two other processes. The process identifier of the activated process buffer with these two arguments would be returned the variable Buff_Pid.

This example also illustrates the use of the Select statement, which provides the mechanism for handling nondeterminism within Concurrent C. A case is selected if the boolean expression (following the keyword case) is true and if a message arrives in its receive clause. If it is possible to select more than one case, then an arbitrary choice is made. When one case is finally selected, the multiple wait terminates and the statements within the case are executed.

Concurrent C provides both message passing and shared variables for interprocess communication. As noted, shared variables are useful only for processes within the same processor. Message passing is done by means of send and receive statements. In a send statement, a sending process must designate the receiving process by specifying its process identifier. A receiving process may or may not designate a sending process. In essence, a sender must know to which process it wishes to send a message whereas a receiver need not necessarily be aware of its sender.

In the statement

```
receive (msg[in++%N]) from sender
```

in the above example, the buffer process can receive a character from the process specified by sender and store it in the character array msg.

Similarly, in the statements

```
receive () from receiver;
send (msg[out++%N]) to receiver;
```

the buffer process is first send a null message from the receiver process and then the buffer process responds by sending the next character to that process. The code within the receiving process might look like:

```
process pl(. . .,Msg_buffer);
  /* the buffer process has been defined by */
  /* a parent process and the process identifier */
  /* has been stored in Msg_buffer. */
  . . .

  send () to Msg_buffer; /* notify buffer process */
  receive (Next_char) from Msg_buffer
  . . .
```

Concurrent C also provides a *monitor* construct for structuring and managing shared variables between processes on the same processor. The concept is similar to the monitor concept of Concurrent Pascal [Brinch-Hansen 1975].

Cooperating Sequential Processes (CSP) The unit of concurrency in CSP [Hoare 1978] is called a *process*; each process being viewed as a sequentially executable sequence of statements. The declaration and initiation of two processes:

```
[TASK_ONE :: body || TASK_TWO :: body]
```

where "body" represents the declarations and statements of each process.

The execution of this statement causes TASK_ONE and TASK_TWO to be concurrently initiated and requires both to terminate before the next statement can be executed.

Communication between processes in CSP is done by synchronization followed by message passing and is accomplished by input and output commands. For example, in the following process body,

```
TASK_ONE ::
  [V1, V2: INTEGER;
  . . .
  TASK_TWO!V1;
  . . .
  TASK_TWO?V2;
  . . .
  ]
```

the output command TASK_TWO!V1 outputs the value of V1 to TASK_TWO. Since message passing can only be done if the processes have been synchronized, if TASK_ONE reaches the output command before TASK_TWO has reached its input command, it must wait. Conversely, if TASK_TWO has reached its input command, it must wait until TASK_ONE reaches the output command.

Once the communication has been completed, i.e., messages exchanges, then separate concurrent execution can continue. In this example, TASK_ONE then executes additional statements and eventually requests input from TASK_TWO via the input command TASK_TWO?V2.

CSP requires that both the sending and receiving processes identify the receiver and sender respectively, i.e., CSP enforces a two-way naming of processes.

CSP also provides constructs for nondeterministic commands. This is based upon Dijkstra's guarded commands [Dijkstra 1975] construct. The buffer example in the previous section might be done as follows in CSP:

```
BUFFER ::
  MSG : (0..255) CHARACTER;
  IN, OUT : INTEGER; IN := 0; OUT := 0;
  *[
   IN < OUT+255; PROCESS_1?MSG(IN MOD 255)
   -> IN := IN + 1;
   ||
   OUT < IN; PROCESS_2?NEXT()
   -> PROCESS_2!MSG(OUT MOD 255); OUT := OUT + 1
  ]
```

A *guard* is CSP may be a Boolean condition followed by an input command; output commands are not permitted within guards. Following each guard is a sequence of statements to be executed.

In the above there are two guards:

```
IN < OUT+255; PROCESS_1?MSG(IN MOD 255)
```

and

```
OUT < IN; PROCESS_2?NEXT().
```

A guard is true if the Boolean condition is true and the input command is ready. The set of true guards determines the sets of actions eligible for execution. If only one guard is true, then the set of actions following the guard is executed; if more than one guard is true, then one is selected for execution.

If the guard

```
IN < OUT+255; PROCESS_1?MSG(IN MOD 255)
```

is true, then PROCESS_1 has sent a character to the buffer process and that character is stored in the message buffer (MSG) and the counter incremented.

On the other hand, if the other guard

```
OUT < IN; PROCESS_2?NEXT()
```

is true, then PROCESS_2 has sent a signal via NEXT( ) to the buffer process that it is ready to receive the next character. The buffer process then sends the next character.

In this example the "*" preceding the "[" identifies the statement as a *repetitive command*. This causes the statement to be repeatedly executed until none of the guards are true in which case the statement executes as a no-op and execution continues with the next statement.

Without the "*", the command is called an *alternative command*. If none of the guards is true when an alternative is executed then an error occurs and the computation is aborted.

Note that unlike Ada or Concurrent C, both sending and receiving processes must be explicitly identified within CSP.

CSP also provides *monitors* [Brinch-Hansen 1975, Hoare 1974] for providing mutually exclusive access to shared concurrent accessible resources.

5.4.4.2 Other Languages and Approaches

In addition to general purpose languages, such as Ada and Concurrent C, with constructs for concurrent processing, several other languages have been designed and implemented specifically for the implementation of distributed systems or components of distributed systems; most of these are experimental.

The Network Implementation Language (NIL) [Parr and Strom 1983] is a high-level language being used for prototyping communication systems. The goal of the NIL project is to develop a programming language to support system design and implementation portable across different hardware and software execution environments.

Systems developed in NIL are composed of independent layers; each layer interfacing to adjacent layers while hiding the internal data and algorithms constituting its implementation. System functions, such as tasking, dynamic introduction of new code, and binding of ports between process instances are provided directly in NIL; escape to the operating system is unnecessary.

NIL provides constructs which permit representation independent description and manipulation of data. It provides the familiar types such as integer, boolean, etc., and also row, table, message, call_interface (for connections to procedures or sending messages), send_interface (for receiving messages), and component (a dynamically load collection of processes and procedures).

A similar approach has been investigated with the development of PLANET (Programming Language for Networks) [Crookes 1984]. The basic component of a PLANET program is the process and they function similar to Cooperating Sequential Processes (CSP). In general, a PLANET program defines a static, hierarchical system of processes which can be mapped onto a configuration of target processors.

A slightly different approach has been explored in PCL (Process Control Language) [Lesser et al. 1979] which permits a programmer to specify process structure in a non-procedural manner. Process description includes both data and control structure information and can be used by the operating system for configuring process structure, scheduling and communication.

PCL could be implemented as an extension to a host language or as a job control language for controlling user modules and where user's modules could use functions of PCL. This latter approach would allow user's modules to be implemented in a variety of languages.

Another potentially interesting approach has emerged as result of the work on SmallTalk [Goldberg and Robson 1983]. SmallTalk provides an object-oriented programming language and environment based upon a message passing paradigm. The object-oriented framework provides an environment for defining objects and localising operations on those objects. The inheritance mechanism provides an approach for the structured development of different classes of objects and the re-use of existing classes and their operations. The message passing communication mechanism between objects seems natural for a distributed environment. Nevertheless, much work on SmallTalk and similar languages remains.

5.5 SOFTWARE INTEGRATION

In practice, the development of a distributed system requires that software exist on a variety of machines, communicate with software on other machines, and interface to existing software: this is the problem of software integration. Although there are no good solutions to these problems in general (except to start from scratch, which may be prohibitive), some approaches can be used to limit integration problems.

Depending upon the nature and functions of the distributed system being developed, some or all of the following integration problems can arise:

1. System components cannot communicate with each other on the same machine, or more likely, on different machines.
2. There are two or more heterogeneous databases, i.e., databases based upon different data models or different implementations of similar data models.
3. Machines within the distributed environment involve several different operating systems.

These types of integration problems are not solely implementation problems. They can be addressed in nearly all phases of the software life-cycle and can be approached from a number of directions. The following sections examine a number of approaches for dealing with such problems.

5.5.1 Planning for Integration

When developing new software it is evident that how the components must interface must be specified and clearly defined. It should also be evident that this is as applicable with existing software. Early in the definition of system requirements and specification, the developers should identify which software components will be retained and incorporated into the existing distributed system and which will be rewritten. In the case of software to be rewritten, new requirements and specifications will have to be developed. For existing software, the interface between it and other components (new and old) must be clearly specified and designed. During these phases it may become clear that it is either too expensive or difficult to integrate certain software and it may have to be rewritten.

Such approaches are particularly important when dealing with heterogeneous databases and/or operating systems. The use of heterogeneous databases within a distributed system may require additional software for the transfer of data between databases or act as filters or front-ends to databases. Such software must be identified, specified, designed and implemented and must be integrated with other components.

Different operating systems often use different values for different meanings (e.g., different values for end-of-file, process interrupt, etc.). Such differences are clear impediments to file transfer and communication between systems. Again, protocols and/or software filters should be identified as part of the requirements and developed systematically as part of the entire distributed system.

5.5.2 Use of Standards

Clearly, one approach to avoiding integration problems is to avoid heterogeneous databases and operating systems and to rely on a single database system and operating system. For example, UNIX exists on a variety of microcomputers, minis, and mainframes and could function as a "de facto" standard for an operating system in a distributed environment. Similarly, one could use SQL as the query language for the databases within the distributed environment. This would provide one mechanism for integrating different databases.

Standardized communications between systems is one particularly useful approach. The reliance on the ISO/OSI communication standards developed and under development form a sound basis (see also Chapter 6). The implementation of the standards on each system within the network provides a common and uniform communication structure. This can help eliminate difficulties in communications between different operating systems.

Also, recent work by the ISO on "standards" for applications, such as file transfer, message passing, and directory service may go a long way to eliminating communication between different operating systems. Of course, the "standards" must be implemented for each different operating system. However, as vendors accept the "open systems concept" more and more, the vendors may provide software support for such services.

Besides technical standards, the IEEE has been involved in the development of standards defining the environment in which software is developed and tested. Standards currently exist for software quality assurance plans (IEEE-Std-730), configuration management plans (IEEE-Std-828), test documentation (IEEE-Std-829), and software requirement specifications (IEEE-Std-830). Pending projects include standards for software reliability measurement, software verification, and software reviews and audits. These standards provide a good foundation for the stages of the software life-cycle and can help ensure that all factors are taken into consideration during each phase.

Also, joint working groups of the IEEE and ANSI are involved in standards activities defining standards for PASCAL and C. Recently, a committee to look at the possibility of a UNIX standard has been formed. The use of standard languages during implementation means that software is more likely to be portable and compatible across systems; this can help simplify integration problems.

5.5.3 Data Directories

One particularly useful strategy for overcoming many integration problems is the use of data directories—dictionaries of programs, databases, files, services, devices, etc. In terms of design and implementation it can provide a central repository for information on record and file formats, message formats, and documentation on the design and implementation of each program. This information can be helpful in eliminating intermodule dependencies such as hard coded message formats and site dependencies. It can also function as a basis for controlling changes in such definitions, tracking changes, and ensuring consistency of critical definitions.

The data dictionary can also play an important role during operation and maintenance. Information on the location of functions or services (i.e. which host), message format for communication, inputs required, outputs produced, etc., can be made available on line. Information generated during the design and implementation phases is then available during maintenance. More details on the contents and use of data directories can be found in Chapter 4.

5.5.4 Design and Implementation Approaches

Identification of software which must be integrated and identification of potential integration problems during requirements is a good starting point for coping with integration problems. The use of standard communications, languages and operating systems for the implementation of a distributed system can help avoid such problems as well. One can also reduce integration problems by adopting certain design and implementation strategies.

Communication via message passing is essential within a distributed environment. Adherence to a design which requires that any two processes, whether on the same machine or on different machines, communicate via message passing, can:

1. Eliminate side-effects and unexpected process interactions
2. Isolate the internals of each process from others, thereby making it easier to change the way a process operates without affecting others

A similar approach has been explored in SmallTalk [Goldberg and Robson 1983], but within an object-oriented environment. Experiences with SmallTalk suggest that a message passing paradigm is useful in limiting interactions.

The work on SmallTalk also suggests an alternative approach to the design of a distributed system: an object-oriented view. In this framework, for example, a particular database would be viewed as an object and a database query would be sent as a message to the database. Such an approach might be applicable in certain application domains, e.g., remote query of distributed databases, but might be less satisfactory in application areas such as electronic mail and conferencing.

Another strategy for implementing a distributed system, the software bus, also makes use of the message passing paradigm. Here, system modules are viewed as "communicating devices" which communicate with each other via a "software bus." The bus, in reality, is another software module whose function is to oversee the message passing—within a single machine and to a counterpart on other machines. With this approach a module need not specify a particular receiver of a message; rather, it need only place a request for information or service "on the bus" which could be handled by one of possible several other processes.

This can reduce specific interprocess dependencies, but perhaps at the cost of greater inefficiency.

5.6 DEBUGGING AND MAINTENANCE

Debugging and maintaining any large software system is difficult. A distributed system presents additional problems because of the presence of multiple processors and interaction between cooperating processes on different processors.

Concern for debugging and maintenance should begin with the specification phase and continue through design and implementation. Recent work (see below) suggests that tools for debugging and maintenance must be incorporated into the design of a distributed system; it might be useful to include such concerns beginning with system requirements.

As noted in Section 5.3, tools to aid in modeling or verifying components of a distributed system can be extremely beneficial in identifying problems before implementation begins. It is estimated that 50–60% of the errors discovered during testing are attributable to misinterpreted or incorrect specifications and designs [Ramamoorthy et al. 1984] and are typically 100 times more expensive to correct than implementation errors [Boehm 1981]. The effort spent during design and specification to avoid errors is quite worthwhile.

Viewing a distributed system as a collection of cooperating sequential programs allows us to partition the debugging problem into two components: debugging a sequential program and debugging interacting processes. Clearly, the first task is to provide sufficient tools, such as interpreters, interactive symbolic debuggers, test case generators, etc., to aid in debugging individual programs.

To aid in debugging cooperating processes, a number of approaches have been proposed. Process traces [Garcia-Molina et al. 1984, Stankovic 1980, Ziegler 1979, Brinch-Hansen 1973] provide a record of "important events" during the execution of a process. If each process in the distributed system produces a trace, then the programmer has a written record describing what occurred in the system. By collecting and examining the traces the programmer can discover what went wrong.

In order to avoid complicating interprocess communication problems, the trace information for a particular process should be kept on the machine on which that process is executing. The programmer must collect this information from the different machines. In many cases, however, the programmer can be supplied with tools to aid in examining the trace information. In particular, the techniques for managing and querying a distributed database can be used to examine the trace data [Garcia-Molina et al. 1984]. The trace information is essentially a read-only database and, hence, the "trace database" can be managed in a simpler fashion. The definition of the trace information and structure of the database must be considered early on in the development process.

Another approach [Garcia-Molin et al. 1984] proposes a distributed debugging facility: a collection of debugging modules, one at each computer within the distributed system. Each debugging module monitors interprocess communication, can examine processes on its systems and can communicate with other debugging modules. The programmer interacts with the debugging system through a particular debugging module and controls the debugging process through that module. For example, the programmer could instruct each debugger to initiate process tracing of each process on its own system or terminate certain processes. Since such debugging facilities must be carefully integrated with the operating system, communications subsystem and programming languages, intended use of such facilities must be considered from the outset of development.

Chapter 6

Communications

Although the concept of an "open system" comes to computer science from the field of communications, in communications the term has a narrower meaning than in the general context of this book: an "open" system is a system which observes a set of communications rules—protocols—which enable it to communicate freely with other systems obeying the same set of rules.

This distinction is closely related to the ongoing debate as to the scope of communications in an open system environment. Japan's Information Network System project, for example, is based on the premise that the nationwide communications network encompasses not only communications facilities, but also information services provided by the network. In the United States, however, such information services are excluded by government regulation from the services provided by communications common carriers. This issue is discussed in relation to the Integrated Services Digital Network (ISDN).

Whatever the approach, if computer systems from different manufacturers are to be open to one another in any degree, the manufacturers must first reach agreement on the protocols to be used in communication. Through such agreements communications standards gradually emerge. A landmark in this field, because of the breadth of its influence, is the derivation by the International Standards Organization (ISO) of a general architecture for the development of standard communication protocols: the Reference Model for Open Systems Interconnection (OSI, often ISO/OSI).

This chapter treats three main subject areas. The first of these is the notion of open systems in communications. An explanation of the open systems concept is followed by a discussion of the ways standards emerge, then by a description of the proposed ISDN standards and the ISDN concept. ISDN is the means by which the carriers attempt to solve the problem of open systems in communications: providing users with a standard digital interface to a large variety of communication services; hiding from the user some of the complexity involved in networking. The networks which ISDN provides access to are built according to the blueprint provided by the ISO/OSI reference model. The implication is that components within ISDN are isolated, in the sense that if one component needs to be replaced (e.g., due to technology changes), other components will be only minimally affected.

ISO/OSI has three levels of abstraction. The OSI reference model is the highest level. A set of service specifications define the reference model in more detail. The least abstract are ISO/OSI protocols; they provide the level of detail required of open systems in order to communicate.

ISO/OSI is gaining widespread acceptance. In a survey conducted in 1986 by the Canadian Federal Department of Communications among computer suppliers, common carriers, and communication users from the government and private sectors, 91% of the 39 respondents agreed that OSI will have apparent impact on the competitiveness of the Canadian industry in the near term (0–5 years), whereas all agreed that the impact will be highly significant in the longer term (6–12 years). Over 90% believed that OSI will provide users with greater choice of software and hardware suppliers; all agreed that OSI will contribute to more universal accessibility of information systems and services.

The move to open systems is having a major impact on communications technology, an impact which will affect the rest of the computer industry. In anticipation of the high volume of data which will flow among systems, and between systems and terminals, ever increasing bandwidth of communication lines is being sought. With fiber optic technology rates of 565 megabit per second (mbps) are currently available, and rates of 1.7 gigabit per second have been announced by AT&T for 1987. In most industrialized countries fiber optic networks are being deployed. In the United States, for example, there are plans to spend about $6 billion over the next five years on fiber optic networks, apart from increased use of satellites and other media. Analysts predict that the result will be a glut in communication capacity. If this indeed will be the outcome, the cost of communication will drop, and a trend will develop to simplify the end systems wherever possible, taking advantage of the cheap and fast communications.

The current state of the art in communications technology is the second subject covered in this chapter. Significant developments which take place within the scope of ISO/OSI are highlighted. Areas covered include local area networking, long-haul communications, and the carriers. Important means of communications which are used to link networks, or which fall outside the scope of local area and long haul networks, are presented in the section on network interconnection.

Finally, this chapter contains a discussion of some important practical aspects, including security, performance, and network management. The issue of system security is important even today, with computer hackers known to have gained access to military, government and other sensitive systems. This problem will intensify with the move to open systems, with information being not only stored in systems but also transmitted over communication channels using standard, public domain communication protocols.

With open systems, network management will become more complex. Given a large variety of options, networks will need to be dynamically configured, automatically tested for correctness, and monitored. Monitoring of the network and its usage will be important for the purpose of billing given that the user will be allowed access to remote systems. Network monitoring will also be used for debugging, tuning, and predicting network performance.

6.1 OPEN SYSTEMS ISSUES

The largest and most sophisticated of today's networks is the public telephone system. It is worldwide, flexible, and enormous—there are more than 500 million telephones in the world. It is also an instance of an open system: from each telephone it is possible to call any other of the 500 million. To make such a call all the user must know is the number of the phone with which he wishes to communicate. He need have no knowledge of the kind of telephone equipment at the other end of the line, or the nature of the links and switches along the way. And once a connection is established, the language and content of the conversation between the two parties is unrestricted.

Any open communications system must provide such "transparency." The means of communication must be so clearly defined that any computer system which complies with a standard interface can connect freely and otherwise transparently to it. As in the case of the telephone system, the user need supply only the network address of the party with whom he wishes to communicate; from that point on the means by which information is transferred between the two points is no concern of his.

Distinctions must be drawn among the means, the context, and the content of communications. For a telephone conversation, there are distinctions between, on one hand, setting up a telephone connection and transferring the voice from one phone to another, and on the other hand, a conversation between partners in some mutually agreeable context—e.g., a language which both partners understand. The content of this conversation—the matter discussed—is yet another matter. These three issues are independent of one another. A telephone connection (means) may carry conversations in various languages (English, Hebrew, etc.—the context); any of a multitude of topics (content) can be discussed in the language selected.

In computer communications these distinctions are equally valid. The networking aspect of a communications system is responsible for setting up connections and handling the transfer of data between connected points. Once a connection has been established various application services (context) may be invoked—e.g., file transfer between machines, or teleconferencing. Finally, once an application service is established, the content—e.g., a specific file to be transferred, a set of messages constituting a teleconference—can be introduced.

6.1.1 Network Directories

A key to the success of any open system is the effective use of network directories. In the case of the telephone system a network directory provides a mapping from the name and geographical address of an individual subscriber to his telephone number. It also includes other information, such as area codes, related to setting up connections.

In an open computer communications system similar information must be available: normally through an on-line service which provides mappings between the name of a communications "target" and its network address. Since it may be possible to reach a particular communications target through more than one network, directory information concerning the target may be quite complex. It should specify the types of networks through which the target can be reached (telephone, telex, etc.), a network address for each of these (telex address, telephone number), and information about application services at the target: for example, type of database, procedures for accessing the database, database subject matter.

Another type of directory information useful in an open system is the "Yellow Pages," which can provide a mapping between a subject (or service) and a set of providers. For example, a medical researcher at a terminal may want information on a certain disease. By means of the yellow pages aspect of the directory system he can discover which databases provide information on that disease. He can then find out how to establish connections with these databases, and how to extract information from them.

With the gradual merging of telecommunications and computer technology, network directories will be supplemented by additional services, mostly related to translation and conversion. In the not too distant future users will be able to use a single standard interface to gain access to many different remote information sources. Another possibility is that users will be able to select their own interfaces, appropriate to their own levels of expertise and professional requirements. In the long run, the advance of "expert" systems will permit users to communicate in natural language with remote sources of information. Systems of this kind, embodying artificial intelligence, may eventually infiltrate the context level of telephone conversations, simultaneously translating telephone conversations between participants speaking different languages. Chapter 4 covers directory services issues in some detail.

6.1.2 Standards

More than in any other area of computer science, standards are crucial in communications. All communicating parties must observe agreed sets of protocols. Agreements on standard protocols emerge in various ways.

They may be established by special standards organizations. Among the most important of such groups are the International Standards Organization (ISO), which sets international standards; the Consultative Committee for International Telephony and Telegraphy (CCITT), an international organization of public telecommunications carriers; professional organizations such as the Institute of Electrical and Electronic Engineers (IEEE) and the Electronic Industries Association (EIA); national or continental organizations like the U.S. National Bureau of Standards and the European Computer Manufacturers Organization (ECMA).

Standards organizations often take the work produced by other standard organizations, adopting the other organization's standard or modifying it. For example, ISO adopting the IEEE 802 set of local area network standards, or the current specification of network directory systems by ISO, CCITT, and ECMA.

Typical standards organizations combine representatives of many interests (e.g., manufacturers, government, user groups, carriers). Because the interests involved are so varied, the definition of a standard normally entails a great deal of pushing and pulling—much time-consuming deliberation and negotiation. Along the way *de facto* standards tend to emerge.

Some products become *de facto* standards by sheer widespread use and adoption by various manufacturers. Two cases in point are AT&T's modems 103 and 212, and IBM's Binary Synchronous Communications (BSC) protocol. AT&T's modems are dominant in the North American market. Being the largest carrier, they determined modem specifications which became a *de facto* standard for using their voice lines for data transmission. BSC, IBM's first commercial synchronous protocol for handling communications between terminal cluster controllers and a host, achieved the status of a *de facto* standard by the volume of IBM equipment which conforms to it. Any other vendor who wishes to communicate with IBM systems or terminals which use BSC protocol, must conform to it.

There's only one IBM. Other vendors who wish to establish standards must ordinarily join forces for that purpose. Xerox, Intel, and DEC cooperated in the specification of the Ethernet LAN product. Their specification later served as the basis of the CSMA/CD network access protocol: IEEE Standard 802.3 and ISO Standard 8802/3. Another case is Intel's Multibus II, which was specified in consultation with 18 other companies.

Another approach open to vendors is to publish their specifications, in hopes that other manufacturers will produce compatible products. Examples are Datapoint's ARC local area network, and Northern Telecom's Open World.

Some customers, like IBM, carry a great deal of clout; an example is the U.S. Department of Defence. When such a customer demands that all their suppliers conform to a specified set of rules, those rules may well become a *de facto* standard. The Defence Department has insisted on COBOL in the past and on Ada at present as the programming languages for their software development. Similarly, the set of Transmission Control Protocol and Internet Protocol became a *de facto* standard because their use was mandated within the Department of Defence. Other instances of customer clout are: (1) Boeing's insistence that IBM provide them with an SNA interface which is compatible with ISO/OSI; IBM has started producing OSI compatible products; and (2) the grouping of users headed by General Motors, which specified the Manufacturers Automation Protocol set for factory automation to which a large number of vendors (e.g., DEC, Honeywell, and IBM) comply.

An interesting development in the standards arena are the efforts extended in encouraging and facilitating the adoption of non-proprietary open system network architectures by both user and vendor communities. A case in point is the Corporation for Open Systems (COS) whose objective is to accelerate the introduction of interoperable, multi-vendor products and services operating under OSI, ISDN, and related international standards to assume widespread customer acceptance of an open network architecture in world markets. Located in Washington, D.C., COS members include users (e.g., Boeing and General Motors), vendors (e.g., IBM, UNISYS, and DEC), and carriers (e.g., Northern Telecom and AT&T).

COS intends to operate through its member companies in pursuing the following strategies:

- To coordinate member companies efforts in OSI, ISDN standard development, protocol selection, conformance testing, and certification
- To work through established standard bodies to expedite the development of OSI and related standards

- To establish a single, consistent set of test methods, test beds, and certification procedures for world markets

6.2 INTEGRATED SERVICES DIGITAL NETWORK

The Integrated Services Digital Network (ISDN) is an architecture supporting digitized telecommunication services which can be selected through a common access-point at a standard interface. In this section the technical aspects of ISDN are discussed, as well as how it will appear to the user, and some of its political and economical ramifications.

6.2.1 Technical Attributes of ISDN

The essential attributes of the ISDN architecture are:

- End-to-end digital connectivity
- Integrated network access
- Out-of-band digital signalling
- Customer control
- A small number of standard interfaces

An explanation of the existing telecommunications system will make it easier to understand what is involved in providing end-to-end digital connections. Historically, analog transmission has dominated the telecommunication industry. In analog transmission one or more of the physical characteristics (e.g., frequency) of a base carrier signal is continuously varied, or modulated, as a function of time.

The present telecommunications network, which was first developed to provide the telephone service, is organized as a highly redundant multilevel hierarchy of switches. Leading out of each telephone are two copper wires that connect it directly to the telephone company's nearest end office. This two-wire connection, between each subscriber's telephone and the end office, is known as the local loop.

The end office can provide a direct electrical connection between two local loops attached to it. In addition, the end office has a number of outgoing lines to one or more nearby switching centers. There are sectional and regional switching centers that form a network; these communicate with one another via high-bandwidth lines called intertoll trunks. The number of different kinds of switching center, as well as the switching topologies, varies from country to country.

To reduce the cost of installing and maintaining lines between switching centers the telephone companies have developed elaborate schemes for multiplexing many conversations over a single physical line. The basic

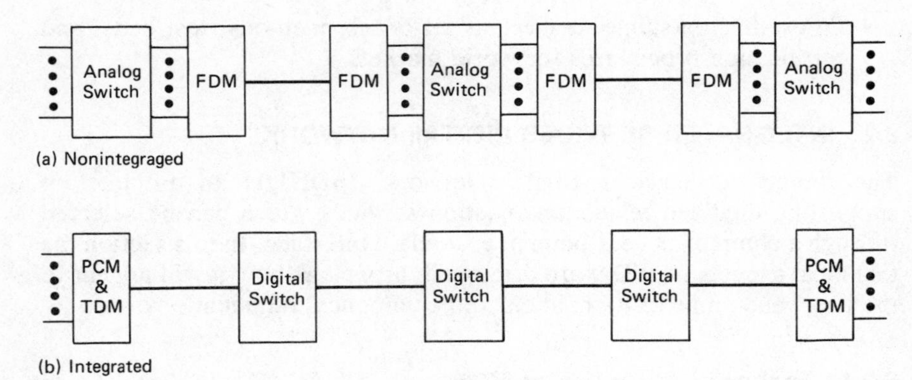

Figure 6.1. The integration of transmission and switching. (*Source*: W. Stallings, "The Integrated Services Digital Network," *Datamation*, Dec. 1, 1984, p.70)

multiplexing category used in analog transmission is Frequency Division Multiplexing (FDM), whereby the frequency spectrum is divided up into frequency bands, with each pair of users having exclusive usage of a frequency band.

With the advent of digital electronics and computers, the switching functions are carried out more effectively by digital computers. Although in industrialized countries high-speed intertoll trunks are being converted to digital transmission, the majority of transmission is still done in analog form. The result is that digital signals are used in switching while analog signals are the primary means of transmission. A typical current scenario is that incoming voice lines are modulated and multiplexed at the telephone company's end-office, then sent over a frequency division multiplexed (FDM) line. At each switching center through which the signals pass, the FDM carrier must be demultiplexed and demodulated before being switched. After switching the signals are once again multiplexed and modulated prior to retransmission.

End-to-end digital connectivity implies the integration of the transmission and switching functions (Figure 6-1). Incoming voice signals are digitized by pulse code modulation (PCM), and multiplexed by means of time-division multiplexing (TDM). In time-division multiplexing the users take turns, each pair periodically getting the entire bandwidth for a small burst of time. Time-division digital switches can switch individual signals without decoding them, thus eliminating the need for separate channel banks which provide multiplexing/demultiplexing at the intermediate switching centers.

Initially the digitization of the telephone network was motivated by the promise of improved transmission quality and reduced overall network costs. It was assumed that there would first be a transition to an integrated

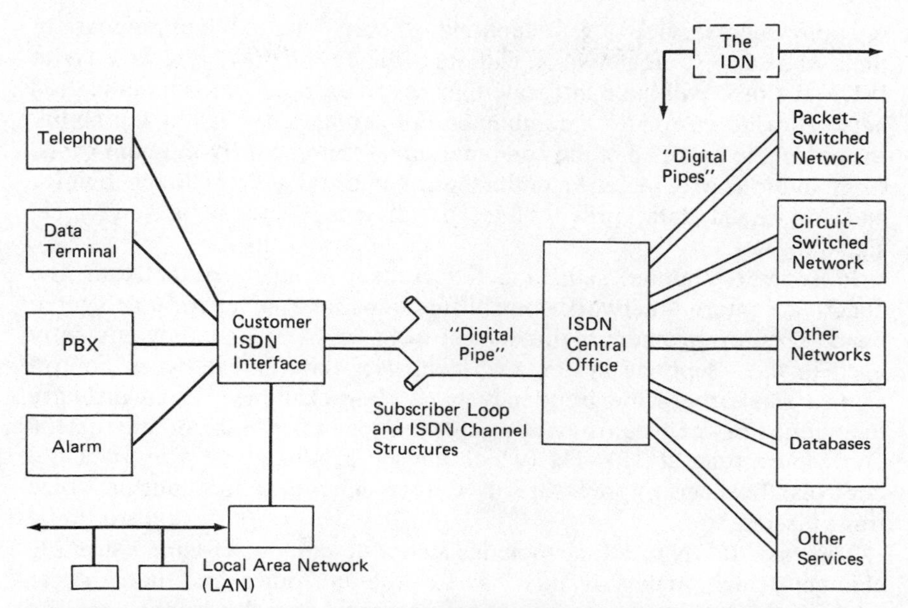

Figure 6.2. Conceptual view of ISDN connection features. (*Source*: W. Stallings, "The Integrated Services Digital Network," *Datamation*, Dec. 1, 1984, p. 74)

digital network, and then a transition to a network which integrated voice and a variety of data services: ISDN. The word "integrated" thus has been used in relation to two quite different concepts.

The main motives for the ISDN are economy of scale and operational convenience. Studies have shown that the marginal cost of carrying data services on the huge network base formed on the digital telephone network will be less than the cost of providing separate network facilities for data services [Gimpleson 1985]. The majority of data services can be integrated into the digital telephone base, forming a general digital transport network for carrying any digital signal (Figure 6-2).

In providing end-to-end digital connectivity, the local loop carries a standard ISDN subscriber package, called the basic interface, which includes two 64 kilobit-per-second channels plus a single 16 kbps channel. The basic interface can be provided on existing twisted-pair loops. In the current technology, 14 kbps full-duplex is considered the limit, even for the most up to date modems, over a voice channel. Moreover, only a single channel is available on the local loop, which requires separate voice and data facilities with separate means of entry for the user. The simultaneous availability to the user of three channels is an important element in the economy of ISDN.

Today each service (e.g., telephone, packet switched data) appears to the user as a separate network, with its own lines and user interface. With ISDN the user will have integrated access to a single telecommunication network which provides a combination of services, as well as the ability to control the make up of the combination on a moment-by-moment basis. Combinations may include simultaneous voice and data, high-speed voice and low-speed data, private line, or all these with separate-channel signalling.

In separate-channel signalling the control signals that activate and deactivate internal network control functions lie in a separate dedicated band. This arrangement is in contrast to the in-band signalling currently used in the telephone system, wherein user data and network control signals share the same frequency band. For example, on an ordinary telephone line the frequency of signals used is in the range of 300 to 3100 Hz; a pure tone at 2100 Hz (which can be produced by a toy whistle) controls the echo suppressors used to eliminate echoes during voice transmission.

The basic ISDN interface includes an end-to-end out-of-band signalling channel which makes it possible for the customer and network to exchange signalling messages. A new protocol is being defined for the signalling channel to provide traffic administration for other channels, billing administration, and customer configured networks. Other functions to be included in the signalling channel may include credit card verification, reception of the caller's telephone number at the called station, flexible call forwarding (based on time or caller), configured billing at either end, and data rates configured by users.

To provide such new services in the telecommunication network it is necessary to pass information among various functional nodes in the network. This could include address information, call control information, or network status information. A separate network, Common Channel Signalling (CCS), is used to provide these functions. A new protocol, Signalling System No. 7 (SS7), was defined in February 1985 by CCITT for the CCS network.

SS7 is a layered protocol which uses high-bit-rate links (56 or 64 kbps), permits long messages (up to 256 bytes of information), and routes information on the basis of its destination address. The new protocol allows for efficient implementation of customer services because of its high message length, and for the transfer of large amounts of data by means of segmentation. The layered structure allows the definition of customer services and the addition of functions as needs arise.

A major achievement of the CCITT in the ISDN area is the development of ISDN user/network interfaces. The importance of these interfaces lies in the need for isolation. The subscriber's terminal equipment must be isolated from internal network configurations and technologies, in order to permit the independent evolution of both. Such isolation will also permit terminal portability among public networks, and between public networks and PBXs.

One requirement of the interface definition was that partitioning of bandwidth and signalling across the interfaces must be flexible in order to meet future service demands.

Two types of ISDN interfaces have been defined:

1. The basic interface, which provides a composite rate of 144 kbps divided into two 64 kbps channels and one 16 kbps channel. (A 64 kbps channel supports the current requirement for digitized voice.) The 16 kbps channel will be used for signalling and for low data rate. Framing overhead, control, and spare bits are additional to the three channels, bringing the actual bit rate at the interface to 192 kbps.
2. The primary interface providing 1.544 megabits per second (in North America and Japan) and 2.048 megabits per second (in Europe) is intended for interconnection with PBXs.

A goal of the CCITT is to define as few interfaces as possible: the smallest number that will economically support a wide range of applications and terminal equipment types. Differences between the telecommunication industry in various countries have had a strong effect on the structure of the ISDN interfaces. See Figure 6-3 for the CCITT architectural reference model describing ISDN interface functions, groupings, and locations. The following definitions are required:

NT1 is network termination device type 1, which contains functions associated with the physical and electrical termination of the network. It provides the user with a standardized interface to the carrier's ISDN and a boundary for the carrier network's local-loop facility. The NT1 device isolates the user from the ISDN local loop transmission technology, which can range from pure digital to existing analog with additional capabilities for digital information. This device will support a new standardized connector for user equipment interfacing the ISDN. Both single and multiple terminals can be supported by NT1. Multiple terminal support is via a passive bus configuration, where contention arbitration on the bus is done at the physical level.

NT2 is intelligent network termination device type 2 which is capable of supporting many S interface connections. The device can be a hub controller (e.g., data terminal controller), a PBX, or a local network. In the latter case, a terminal with the ISDN interface S would connect to a

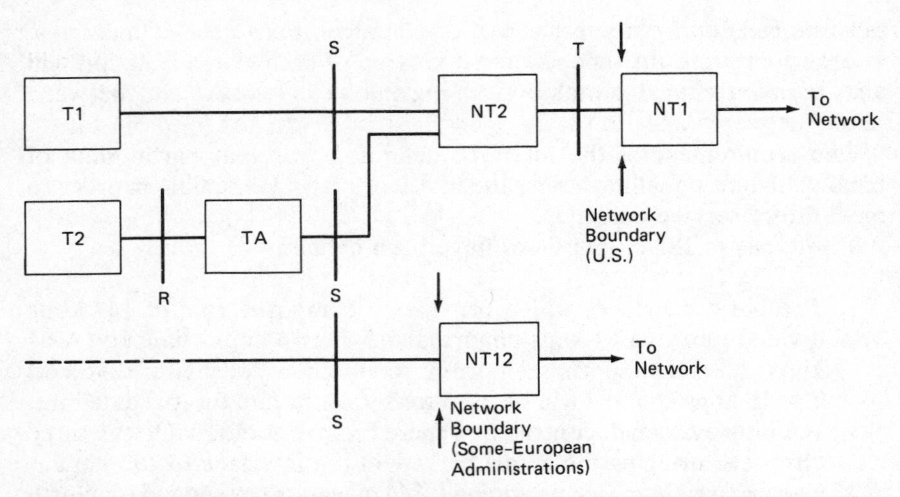

Figure 6.3. ISDN simplified. R, S, and T are interfaces at the referenced locations. NT1, NT2, NT12, T1, T2, and TA are functional groupings. R represents an existing interface, such as RS-232-C, while S and T are ISDN interfaces. T1 and T2 are terminals; TA, a terminal adapter; NT1, a network terminator; and NT2 and NT12, CBXs. (*Source*: B. E. Collie, L. S. Kayser, and A. M. Rybczynski, "Looking at the ISDN Interfaces. Issues and Answers," *The Executive Guide to Data Communications*, Vol. 7, p. 43)

media access unit which resides on the local network. NT2 functions may include switching, data concentration, multiplexing, protocol handling and maintenance at the data-link and network levels.

NT12 is a single device which contains the combined functions of NT1 and NT2.

TA is a terminal adapter which translates existing interfaces, e.g., X.21 or RS-232-C, into the ISDN interface.

T1 is terminal type 1 which supports the ISDN interface.

T2 is terminal type 2 which requires a terminal adapter (TA) interface in order to plug into an ISDN interface.

A key reason for the separation of the NT1 functions from NT12—and the resulting creation of two types of network termination—is that in most countries the state (Postal, Telephone, and Telegraph or PTT) supplies communication services, and the network termination device may contain the NT12 functions. In the U.S. there is a need for network termination with a limited number of functions, since the Federal Communications

Commission (FCC) has deregulated and unbundled the equipment on customers' premises from the offerings of regulated carriers (for example, AT&T is prohibited from offering an RS-232-C interface to its Dataphone Digital Service). From the U.S. regulated carriers point of view, NT1 should provide adequate separation and control for network termination, but its intelligence should be minimal; otherwise its cost may result in pressure from users for NT1 ownership.

6.2.2 User Perspectives on ISDN

The standard interface provided to the ISDN user will allow entry into the network by means of a wide range of terminal equipment, including conventional telephones, terminals, and integrated telephone/terminal units. The introduction of microchips providing standard ISDN interface functions will result in a reduction in price and an increase in variety of the terminal equipment. Motorola, for example, has announced a chip in the 145420 series which supports the physical level functions of the ISDN basic interface.

A standard interface allows the user to plug in a terminal to any access point of the network. Terminals should be almost as portable as telephones in todays telephone network. It is envisioned that intelligent service nodes in the ISDN network will maintain customer profiles; when the customer moves from place to place, his unique profile will be moved to the new location.

ISDN will provide many new services, some which are currently provided as overlays on existing (largely analog) networks. Candidate services for integration in the ISDN are listed in Table 6-1.

From the user's perspective, the two 64 kbps basic interface channels will provide him with excellent quality of voice and data service. The 16 kbps channel provided in the basic interface will not be dedicated to network signalling but could be used to provide services such as telemetry and low-speed data transmission. The services provided will be enriched with a plethora of features, some of which are listed in Table 6-2.

It is expected that the transition of the telephone network from analog to digital will take from 10 to 20 years. Transition plans vary from country to country. During the transition period the current analog systems and the new digital systems will coexist on the same physical cable systems. At first, analog/digital converters will be needed to interconnect the two types of exchanges. Gradually, however, the goal of an all-digital network will be achieved. Three implementation phases for subscriber services are planned: narrowband, wideband, and broadband.

The narrowband phase implements the basic and primary interfaces described above. The wideband ISDN uses bit-rates from 64 kbps to a

Table 6.1. Candidate Services for Integration.

BANDWIDTH	SERVICE			
	TELEPHONE	DATA	TEXT	IMAGE
Digital voice (64 kbps)	Telephone Leased circuits Information retrieval (by voice analysis and synthesis)	Packet switched data Circuit switched data Leased circuits Telemetry Funds transfer Information retrieval Mailbox Electronic mail Alarms	Telex Teletex Leased circuits Videotex Information retrieval Mailbox Electronic mail	Videotex Facsimile Information retrieval Surveillance
Wideband (>64 kbps)	Music	High-speed computer communications	Teletex	Teletex Tv conferencing Videophone Cable tv distribution

SOURCE: W. Stallings, "The Integrated Services Digital Network," *Datamation*, Dec. 1, 1984, p. 79.

Table 6.2. Basic and Additional Facilities for ISDN Services.

BASIC FACILITIES

TELEPHONY	DATA	TELETEX	VIDEOTEX	FACSIMILE
National toll access	Automatic dialed call	Incoming call not disturbing local mode	Information retrieval by dialog with a database	Automatic dialed call
International toll access	Manual dialed call	Message printed on operator demand		Manual dialed call
Malicious call blocking	Automatic answer	Message presentation as in the original		Automatic answer
		Day and hour automatic indication		

ADDITIONAL FACILITIES

TELEPHONY	DATA	TELETEX	VIDEOTEX	FACSIMILE
Transfer call	Direct call	Delayed messages	Transactions	Delayed delivery
Abbreviated dialing	Closed user group	Abbreviated address	Message box service	Multiple destination
Rerouting to verbal announcements	Closed user group with outgoing access	Multiple address	Loading of software from a database to a terminal	Code, speed, and format conversion
Intermediate call	Calling line identification	Charging indication	Loading of special character set	
Conference call	Called line identification	Telex access		
Camp on busy	Abbreviated address calling	Graphic mode		
Barring outgoing toll traffic	Barring incoming call			
Hot line	Multi-address calling			
Detailed billing	Detailed billing			
Automatic wakeup	Transfer call			
	Call charging indication			

SOURCE: W. Stallings, "The Integrated Services Digital Network," *Datamation*, Dec. 1, 1984, p. 80.

maximum of 2.048 mbps. The wideband services will be provided as transparent circuit-switched channels at bit-rates selected by the subscriber. Subscriber connections in local exchanges may range from twisted-pair wires for connecting relatively slow services to coaxial cables or optical fiber lines for high-rate services. The wideband switching mechanism must be programmable to provide any bit-rate from 64 kbps to 2.048 mbps, in 64 kbps increments.

For speech, text, most data, and still-picture transmission, the 64 kbps rate is sufficient. The wideband ISDN will be able to handle speech transmission with radio quality, as well as high-resolution moving pictures. When it comes to transmitting at the exceptionally high data-rates needed for multiple-channel high-resolution television, or for quick mass file transfer, even coaxial cable is no longer sufficient. In the third phase of ISDN, optical fiber technology will be used to provide transmission speeds as high as 140 mbps, thus permitting the integration on a single line of all the services described above.

Because ISDN is a system for the future, precise definitions for the services to be provided have not yet been formulated. However, whatever systems are used, the services will permit open communications and control; this implies that the systems must be open in relation to new facilities, services, and applications. In addition, they must be applicable to both public and private environments.

6.2.3 Political and Economical Ramifications

A massive standardization effort is under way, under the auspices of the CCITT, with significant work taking place in North America, Europe, and Japan. At the CCITT plenary session held in November of 1984 the entire body adopted a general framework for ISDN standards, including a reference configuration for the user interface.

Within CCITT there are profound differences in opinion between American and European representatives as to what constitutes an ISDN. U.S. representatives envision ISDN as a pipeline, linking any one terminal to any other terminal or computer provided by individual vendors. The approach renders the ISDN a pipe connecting the user to various service centers.

In the eyes of the Europeans, on the other hand, the ISDN should provide both interconnection and capability. The service centers are seen as imbedded in the ISDN. In Europe the various national Postal, Telephone, and Telegraph (PTT) organizations have monopoly control over the communications industry. The PTTs generally own the terminal equipment and information services, and can therefore determine the character and use of the entire system.

In the United States the situation is considerably different, because of the structure of the communications industry following the AT&T divestiture. A communications circuit in the U.S. involves a local exchange carrier or bypass agency, a long-distance or common carrier, and possibly a value-added carrier. An effective ISDN implementation will require the cooperative effort of all these independent agencies.

In the U.S. a national ISDN on the European model is simply illegal. Current regulatory constraints prevent American carriers from offering many of the data processing services conceived by the European planners. It is through the efforts of the American representatives that NT1 and NT2 have been separated in the CCITT reference configuration. Under current U.S. regulations NT1 is part of the carrier's network, while NT2 is considered Customer Premises Equipment (CPE). Until recently the major carriers were forbidden to provide NT2, except through separate and independent subsidiaries.

There are other political ramifications. For example, in most countries there are separate networks for voice and data, both owned by the national government. The government can achieve both greater efficiency and greater control by integrating these services in a single network. ISDN, with its potential for new services, is seen as a path toward an "Information Society," in which data can be shared throughout the nation over the network. But then, such a network might well increase the level of government access to information pertaining to both individuals and organizations, and therefore gives rise to fears for privacy: there are differing views as to the status of information as a public resource.

The flow of information across national boundaries, in an international ISDN, is another issue. Even if all the countries involved adopt the same set of communications standards, there will be the problem of differing laws governing the movement of information within particular countries and across international borders.

Finally, there are the human concerns. The ISDN will constitute a major technological revolution, and its introduction will inevitably affect the security and stability of jobs. Trade unions, especially in countries which have national PTTs, are taking a close look at the likely implications.

The major forces pushing for ISDN were initially the Europeans, who needed to upgrade their existing telecommunication networks. The European push has slowed down due to economic constraints, and it seems that current ISDN impetus is provided by the U.S. and Japan who see large potential markets in ISDN technology. Almost all U.S. telephone companies are currently testing ISDN technology for data and voice integration (terminals, local loops, etc.).

On the other hand, there are economic constraints on the Americans as well. ISDN entails the replacement of the existing telephone network, and in North America that network is quite adequate to the current demands of its users. It is, moreover, an enormous investment; it is designed to pay for itself over a long period, and cannot economically be replaced before the end of that period. Countries with well developed economies and aging telecommunications systems, such as Italy and France, are better placed in this respect, and will likely be the first to undertake major replacements.

6.3 ISO STANDARDS FOR OPEN SYSTEMS INTERCONNECTION

6.3.1 Introduction

In March of 1978 the first meeting of the ISO Subcommittee on Open Systems Interconnection (Technical Committee 97, Subcommittee 16) met in Washington, D.C. Its task was to develop a reference model for a general architecture which might serve as a context for the development of worldwide standards for distributed information systems.

Once the reference model was established, the task of the subcommittee would be to analyze existing ISO standards to determine whether or not they meet the general criteria of the model, and if not, what changes will be required to bring them into alignment with it. Areas where no standards existed were also to be identified for future development. The OSI Reference Model was therefore to be a framework in which the development of all communications standards can be coordinated.

The term "open" was chosen to emphasize that if it conforms to the OSI (Open Systems Interconnection) standard a system will be open to communication with any other system obeying the same rules.

By the spring of 1983 the Basic Reference Model had been formulated, and was adopted as an international standard [ISO 7498 1983].

The problem faced by the ISO subcommittee was considerable: they were asked to develop a set of general standards on which emerging products might be based, *before* commercial practices in distributed systems were clearly established, and when many fundamental technical problems remained unsolved. The subcommittee has dealt with this predicament in such a way as to maximize the flexibility of the Basic Reference Model and to minimize the impact of technological change on that model.

Their approach has been to define the issues related to communications among systems in terms of a layered architecture, each layer representing a manageable piece of the whole. Communications problems are approached in a top-down manner, successively embodying three levels of abstraction:

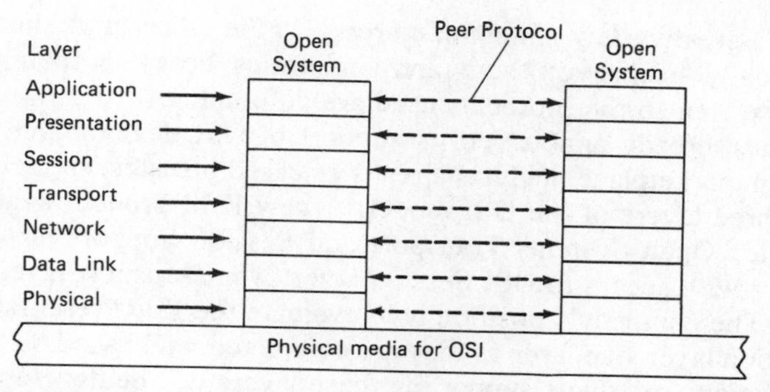

Figure 6.4. Seven layer reference model and peer protocols. (*Source*: ISO 7498 1983)

1. The architecture
2. The service specification
3. The protocol specification

The architecture is described in the Reference Model, ISO 7498. The reference model defines the types of objects used to describe an open system, the general relationships among such objects, and the general constraints on both objects and relationships. It also elaborates a seven-layer model, constructed in terms of these objects, relationships and constraints, for interprocess communication (Figure 6-4).

OSI service specifications define in greater detail the service provided by each of the seven layers defined in the reference model. A service specification defines the facilities to be provided to the user of a service, independent of the mechanisms used to provide that service. In addition, it provides an interface for the layer in question by defining functional primitives which may be requested by the user. The service specification is done without regard to implementation, such that different protocols can comply with the same service specification.

OSI protocol specifications impose tighter constraints on service specifications, which have a direct effect on implementation. Each protocol specification defines precisely what control information is to be exchanged, and what procedures are to be used to interpret this information.

In the world of OSI, only OSI protocols are implemented, and all products conform to such protocols. The statement "this product conforms to the OSI Reference Model" does not imply an ability to interwork with other products for which the same claim is made.

Another common concern is the relationship between OSI and IBM's System Network Architecture. SNA has been designed for use in an environment in which all components are produced by IBM. The purpose of OSI is to interlink a heterogeneous environment, in which components are from a great variety of manufacturers. On the other hand, the two are not mutually exclusive; there are similarities between their layered structures, though the protocols used are different.

IBM has already announced its support of OSI development for the European marketplace, and has already released products supporting the lowest three layers of the OSI model. A new IBM product available in 1986 called Open Systems Transport and Session Support provides an IBM user with access to OSI Session layer over Transport layer classes 0 and 2. The company's position is to evaluate the ISO Presentation and Application layer standards as they are completed, and based on business considerations, products supporting these layers will be developed.*

By 1990 SNA and ISO/OSI will certainly coexist; in the longer term they may well converge.

6.3.2 Elements of the OSI Architecture

The elements of the OSI architecture are building blocks used to construct the seven-layer model. In this architecture communication takes place between application processes running in distinct systems. A system, in this context, is one or more autonomous computers (with their associated software, peripherals, and users) which are capable of processing or transferring information.

Layering is a structuring technique which allows the logical decomposition of a network of open systems into smaller subsystems; interfaces among these layers are clearly defined. A layer is composed of subsystems of the same rank in all interconnected systems. Each layer adds value to services provided by the set of lower layers, in such a manner that the highest layer is offered the full set of services required to run distributed applications.

As in structured programming, where only the functions performed by a module, not its internal operations, are known to the users of that module, each OSI layer provides a set of services to the next higher layer, which need not know how these are performed. Each layer is therefore independent, in the sense that changes can be made in its internal mode of operation, provided only that it continues to deliver the same service.

The following notations and definitions will be used to describe OSI layers (Figure 6-5). Any layer in the OSI model is referred to as the (N)-layer. The same notation is used to designate all concepts relating to individual layers: e.g., entities in the (N)-layer are (N)-entities.†

---

*One such product introduced in 1986 is the General Teleprocessing Monitor for OSI. It contains programming tools for writing OSI Presentation and Application protocol conversion routines to support communication between OSI and SNA applications.

†The layer above the (N)-layer is the (N+1)-layer, the layer below is the (N−1)-layer.

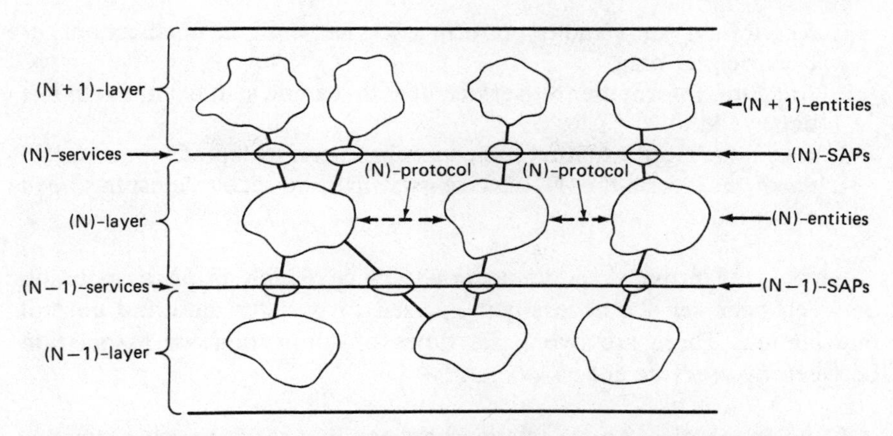

Figure 6.5. Entities, service access points (SAP's), and protocols. (*Source*: Day and Zimmermann, "The OSI Reference Model," *Proceedings of the IEEE*, Vol. 71, No. 12, Dec. 1983, p. 1336)

An (N)-protocol is the set of rules and formats which govern communications between (N)-entities in different open systems.

(N)-services are offered at the (N)-service access points, which are the logical interfaces between the (N)-entities and the (N + 1)-entities. These services are provided by means of primitives which represent input commands to an entity. Four generic primitives have been defined (Figure 6-6):

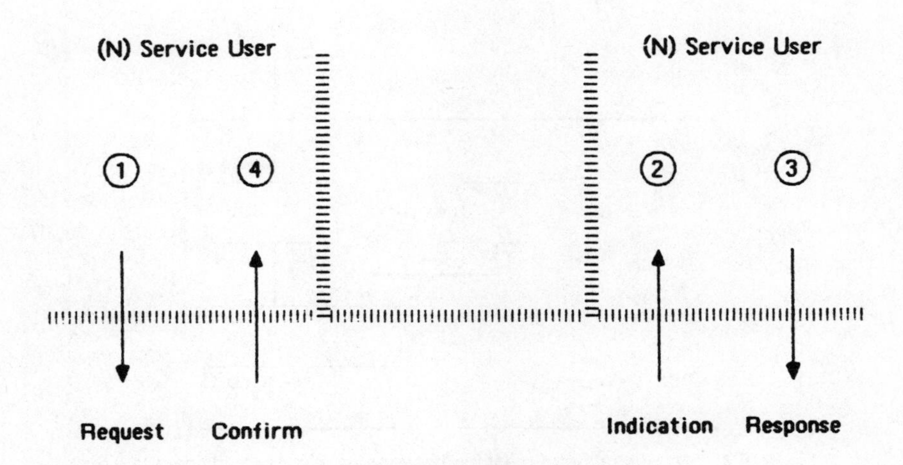

Figure 6.6. Generic service primitives.

1. *Request*. A command to perform a service: e.g., establish a connection, deliver data.
2. *Indicate*. Inform the (N)-service user that some significant event has occurred.
3. *Respond*. Notify that an indication has been accepted.
4. *Confirm*. Inform the (N)-service user that a service request has been executed.

Each layer provides a connection service which is an association between peer service access-points, used to transfer data and control information. There are two basic types of connection, or association between two service access-points.

1. *Point-to-point*. An association between two service access-points.
2. *Multi-point*.An association among multiple service access-points (as in broadcast or multidrop communications).

Information is transferred in units of various types (Figure 6-7):

1. (N)-user-data is the data transferred between (N)-entities on behalf of the (N + 1)-entities for whom the (N)-entities are providing services.
2. (N)-service-data-unit is an amount of (N)-interface-data whose identity is preserved from one end of an (N)-connection to the other.

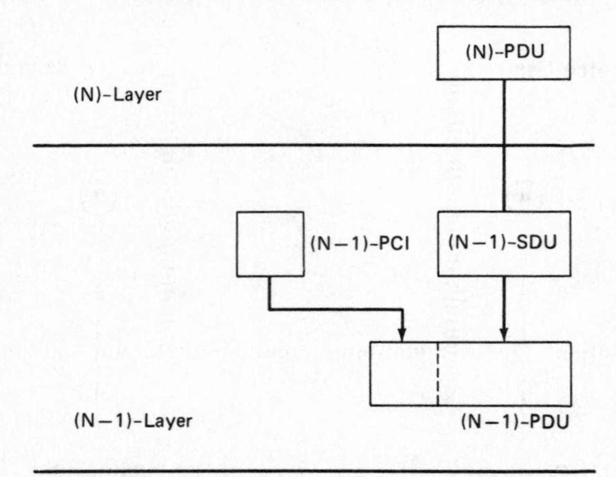

Figure 6.7. An illustration of mapping between data-units in adjacent layers.
PCI = protocol-control information
PDU = protocol-data-unit
SDU = service-data-unit

3. (N)-interface-data is information transferred between an (N + 1)-entity and an (N)-entity, for the purpose of transmission between two (N + 1)-entities over an (N)-connection.
4. (N)-protocol-control-information is information exchanged between two (N)-entities, using an (N − 1)-connection to coordinate their joint operation.
5. (N)-protocol data units combine (N)-protocol control information and, possibly, (N)-user-data, and are exchanged between peer (N)-entities.

Each layer treats the protocol data unit of the layer above as inviolate user-data, not to be altered or contaminated. Layer (N), on receiving an (N + 1)-protocol data unit, will add this unit to its own protocol control information, forming an (N)-protocol data unit to be transferred to layer (N − 1). On the other side, layer (N) will receive from layer (N − 1), an (N)-protocol data unit, strip from it the (N)-protocol control information, and transfer the (N)-user-data to layer (N + 1).

Three types of mapping are defined between (N)-connections and their supporting (N − 1)-connections (Figure 6-8):

1. One-to-one. A single (N)-connection on top of one (N − 1)-connection.
2. Multiplexing. Several (N)-connections on top of a single (N − 1)-connection.
3. Splitting. A single (N)-connection on top of several (N − 1)-connections, dividing traffic among them.

Flow control is observed between peer-entities within the same layer and across the interface between layers. Error functions addressed by the model include acknowledgment, error detection, and error verification.

6.3.3 The Layered OSI Reference Model

In open systems interconnection, interprocess communication is subdivided into seven independent layers. These are described briefly, from the bottom up.

Physical Layer The lowest layer provides the mechanical, electrical, functional, and procedural standards which permit access to the physical transmission medium. Services provided to the Data Link layer by the Physical include the establishment, maintenance, and disconnection of physical connections.

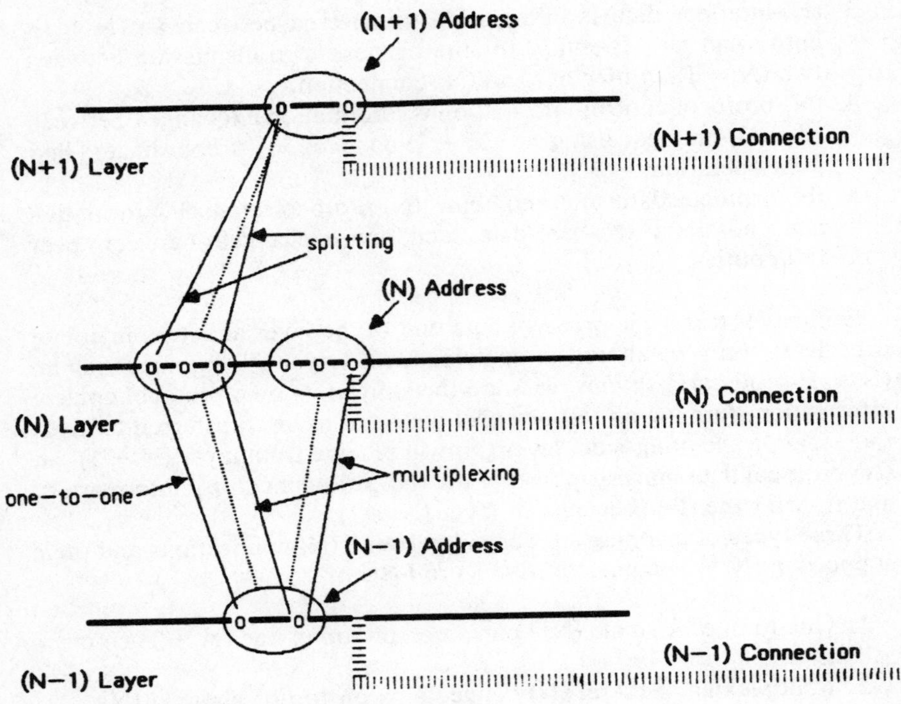

Figure 6.8. Address-mapping among layers.
(N) Address = (N) Address of (N) Service Access Point
o = Connection Endpoint Identifier

Physical layer protocols represent the traditional interfaces between data terminal equipment (DTE) and data communications equipment (DCE). These handle the electrical and mechanical interface requirements, bit synchronization, and identification of the signal elements as either 0 or 1.

For serial transmission, the physical service data unit is a single bit. If the medium supports parallel paths, as on an 8-bit bus, the basic unit is eight bits transmitted in parallel.

Typical Physical layer protocols are RS-232-C and RS-449, issued by the Electronic Industries Association (EIA). Others, such as CCITT X.21 are also widely used.

Data Link Layer This layer provides the means of transferring data between Network layer entities, while detecting and, if necessary, correcting errors which may occur in the Physical layer. Data Link protocols and services are designed in order to ensure effective and efficient use of a wide variety of physical transfer media. Typical Data

Link protocols are the bit-oriented ISO High-Level Data Link Control (HDLC) for point-to-point and multipoint connections, and the character-oriented BSC.

Network Layer The Network layer provides independence from specific data transfer methodologies, as well as from relaying and routing considerations. The basic service provided is the movement of data through a network consisting of concatenated data links, with multiple routes available between points. An internetworking service, routing and relaying information among concatenated networks, may also be provided. The best known Network level protocol is the CCITT X.25, for packet-switched networks.

Transport Layer This layer provides transparent transfer of data between source and destination systems, relieving the higher layers of any concern with reliable and effective data transfer. Transport protocols ensure end-to-end integrity of the data exchange, bridging the gap between services provided by the underlying Network layer and those required by the Session, Presentation, and Application layers. Transport protocols can be simple or complex. Simple protocols are used when the underlying network provides a high-quality, reliable service; more complex protocols must be used when the underlying service is poor, to ensure reliable end-to-end transfer.

Session Layer This layer provides the structure which controls interactions among application processes; it includes mechanisms for providing two-way simultaneous and two-way alternate session connection and release, and for the establishment of major and minor dialogue synchronization points.

Presentation Layer The Presentation layer provides application processes with independence from data syntax. Presentation protocols allow the user to select a presentation context which defines the representation of the data to be transferred. The context may be specific to an application or to a type of hardware, or it may meet some other standard of representation. Presentation layer services include translation, transformation, and formatting according to the selected presentation context.

Application Layer Since it is the highest ISO layer, the Application layer does not provide services to other layers. Application processes reside here: specifically, those aspects of application processes which pertain to interprocess communication. The ISO is currently developing a set of common application service elements, which will provide common procedures for constructing application protocols and for gaining access to OSI services. Three Application protocols of general interest are also under development: for file transfer, access and management, virtual terminal, and job transfer and manipulation.

6.3.4 Advantages and Disadvantages of ISO/OSI

Advantages The OSI Reference Model provides a coherent structure for communications, with respect to standards and protocols. In addition, it offers a blueprint for the specification of standards, in terms of scope, vocabulary, and structure—it is a reference for standards.

The layered structure of the OSI allows flexibility, and ensures the independence of each layer from all others. By hiding the internal operation of the layer—only the interface matters—abstraction is provided to the users. Logical separation of the service provided from the various possible physical means of providing it makes it possible to change implementation of specific functions without altering others. It also makes possible a ''building block'' approach, in which relatively small, clearly defined problems can be attacked independently, isolated from other issues by the OSI structure. Illustrations of this are to be found in the discussion of LANs and of the X.25 packet-switching protocol.

Standard interfaces permit interchangeability of components, allowing the mass production of hardware components. Similarly, a software package can be used by a large number of users. The result is reducing cost and freeing users from dependence on specific vendors, giving a degree of flexibility which would be otherwise unattainable.

Finally, the OSI Reference Model stands a good chance of achieving genuine international acceptance, a rare attainment indeed. If OSI is indeed accepted worldwide, it will open the way to truly open systems.

Disadvantages Overhead is the most obvious disadvantage. OSI standards must attempt to cover many possible different ways of communicating, and therefore must embody functions which are not necessary in particular implementations.

Rigidity is another. This is a problem with all standards: they limit the number of possible approaches to a problem, and may just preclude the development of a valuable product or method. They tend to restrict innovation. Standards can be changed, but normally such change occurs very slowly.

There is a danger that consumers will delay purchases and manufacturers will hold back investments, until both are sure that what they buy or sell conforms to a forthcoming standard.

The field of communications, on the other hand, is relatively young and changes quickly. It is difficult, first, to anticipate the direction of change, and secondly, to keep pace with technological change in the generation of OSI standards.

The whole process of OSI development is very time consuming. Many people are involved, and these people represent many different interests on both the vending and purchasing sides—among the results is a multiplicity of standards. Paradoxically, mostly bigger organizations are currently represented; smaller vendors and users have little say.

6.4 LOCAL AREA NETWORKS

Within the framework of the ISO/OSI reference model there is a vast number of possibilities for connecting systems in a network. The complexity of the problem is reduced by distinguishing networks according to the distance they cover. The main distinction is between local area networks (LANs) and long haul networks.

While there is no cut and dried definition for a local area network, it is possible to list some basic characteristics of LANs, which tend:

1. To cover a distance of no more than a few miles, normally up to 10 miles
2. To be owned and used by a single organization
3. To allow for very high data-rates, normally between 0.1 and 100 megabits per second
4. To have low error-rates, in the order of 10^{-9} (i.e., an average of one bit is in error for every 10^9 bits transmitted)

One way of locating LANs in the order of things is to say that they fill a logical gap between multiprocessing computers and long-haul networks: allowing interconnection of computers in order to gain the resource sharing of computer networking and the parallelism of multiprocessing [Metcalfe and Boggs 1976].

During 1983–84 the net expenditures for LANs were $224 million; it is expected that this figure will rise to $1.31 billion by 1988 [Rosenberg and Feldt 1984, quoting Carol Snell of Dataquest Inc.]. There are several reasons for such a rapid growth in the numbers of LAN installations.

First, they permit the sharing of resources. A LAN can support servers such as a large on-line storage system (a file server), a laser printer, or a long-haul communications server. Through the LAN these servers support a community of users of PCs and terminals which can be connected at will to minis and mainframes.

Second, LANs permit the easy exchange of data among systems. As a general rule for any business organization, more than 75% of all communications are internal. The LAN supports such internal communication effectively.

The discussion in this section concentrates on the LAN 8802 standards defined by the IEEE 802 Technical Committee [ISO 8802 1985], which have now been adopted by the ISO. There are of course many LANs which do not conform to these standards. Some of these were produced before the standards were issued; others are tailored to specific applications.

6.4.1 LAN Technology

Local area network technology will be discussed in terms of three considerations: topology, transmission medium, transmission technique.

Topology IEEE 802 standard LANs come in two topologies, bus (and tree) and ring. In either of these, data is transmitted in the form of "frames," which contain both the source and the destination addresses as well as the message.

Non-standard LANs include star and hybrid topologies (Figure 6-9).

The bus topology is characterized by the use of a multiple-access broadcast medium. Because all devices share a single common medium, only one device can transmit at a time—otherwise messages will collide. Each station on the LAN monitors the medium constantly, and copies message frames addressed to itself. Ethernet is an example of a bus topology LAN.

The tree topology is a modification of bus which can be employed with a broadband coaxial cable, such as LocalNet 20 by Sytek. The unidirectional signal on coaxial cable permits construction of a tree architecture.

The ring topology consists of a closed loop, in which each node is attached to a repeating element. Data circulates around the ring on a series of point-to-point links between repeaters. A station wishing to transmit waits for its turn and then sends a frame onto the ring. As the frame circulates, the destination node copies the data into a local buffer. The frame continues to circulate until it returns to the source node, providing a form of acknowledgment. An example is IBM's ring-based LAN.

A star topology consists of a central switching element which connects all the nodes in the network. A station wishing to transmit data sends a request to the central switch for a connection with some destination station. The central switch establishes a dedicated path between the two stations (by circuit switching). Once the path is established, data can be

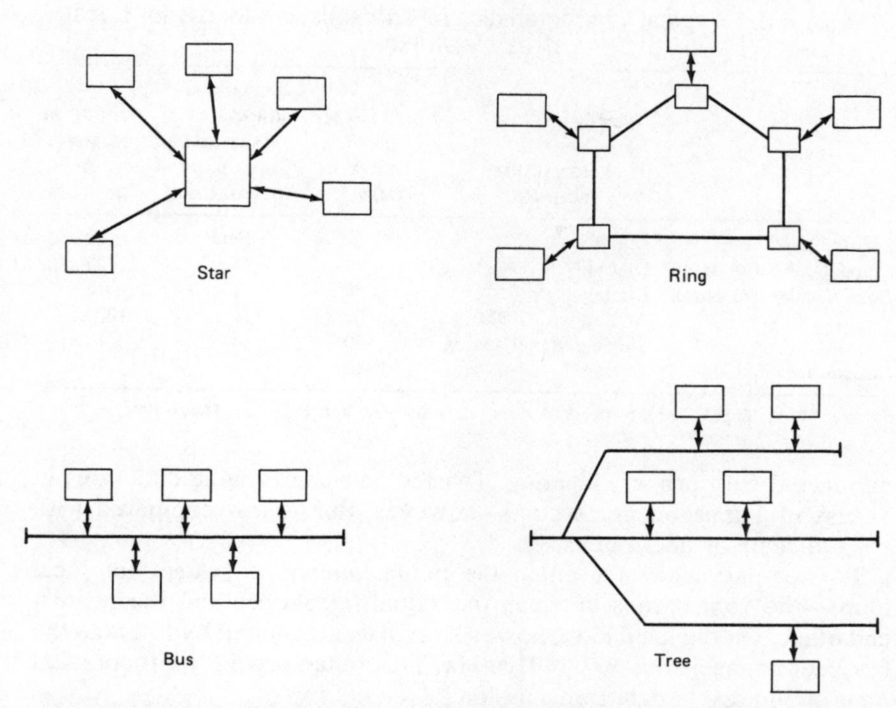

Figure 6.9. Local network topologies. (*Source*: W. Stallings, "Local Networks," *Computing Surveys*, Vol. 16, No. 1, March 1984, p. 6)

exchanged between the two points as if they were connected by a dedicated point-to-point link. The private branch exchange is based on star topology.

Hybrid topologies are simply combinations of these basic forms. For instance, a ring can be combined with a star topology to form a network in which messages between any two nodes can pass either through a central switch or around the ring. An example of a hybrid topology is the CN-III network developed at the Imperial College in London, England [Barnett and Beckwith 1983].

Transmission Media The media most appropriate to LANs are twisted-pair wires, coaxial cable, and optical fiber cable. Table 6-3 compares typical characteristics of these transmission media. Twisted-pair wire is normally a cost-effective choice for low-volume traffic within a single building [Stallings 1984]. Such wiring is often pre-installed in office buildings. A weakness of twisted-pair is its susceptibility to interference and noise, including cross-talk from adjacent wires. These effects can be

Table 6.3. Typical Characteristics of Transmission Media for Local Area Networks.

	SIGNALING TECHNIQUE	MAXIMUM DATA RATE (MBPS)	MAXIMUM RANGE AT MAXIMUM DATA RATE (KILOMETERS)	PRACTICAL NUMBER OF DEVICES
Twisted pair wire	Digital	1–2	Few	10's
Coaxial cable (50 ohm)	Digital	10	Few	100's
Coaxial cable (75 ohm)	Digital	50	1	10's
	Analog with FDM	20	10's	1000's
	Single-channel analog	50	1	10's
Optical fiber	Analog	10	1	10's

SOURCE: W. Stallings, "Local Networks," *Computing Surveys*, Vol. 16, No. 1, March 1984, p. 7.

minimized with proper shielding. Twisted pair can provide data-rates in excess of 1 megabit per second—however, this is low compared with coaxial cable or fiber optics.

Twisted-pair wire is used in the public telephone system for local loops—the connections between individual telephones and the nearest end office. On the local loop, however, signals are limited by filters to the frequency range from 300 to 3100 Hz. This range provides a theoretical limit of 30 kbps in data transmission.

Coaxial cable offers higher performance than twisted-pair wire; it is capable of higher throughput, supports a greater number of devices, and can span greater distances. The two main types of coaxial cable now in use are:

1. 75-ohm, standard in community antenna television (CATV)
2. 50-ohm, used by most baseband coaxial systems

Optical fiber cable has a number of advantages over both twisted-pair and coaxial cable: light weight, small diameter, low susceptibility to noise, and negligible emissions. To this point, however, the use of optical fiber has been limited by cost and by technical limitations. Point-to-point fiber optic technologies, such as ring networks, are becoming feasible as the cost of fiber optic equipment declines. Multipoint topologies like the bus are difficult to implement in optical fiber, because each tap (a station's attachment to the cable) causes significant loss of power, as well as optical reflections.

The choices of transmission medium and topology cannot be made independently. A bus topology can be implemented with either twisted-pair or coaxial cable. Twisted-pair, coax, or optical fiber can all be used

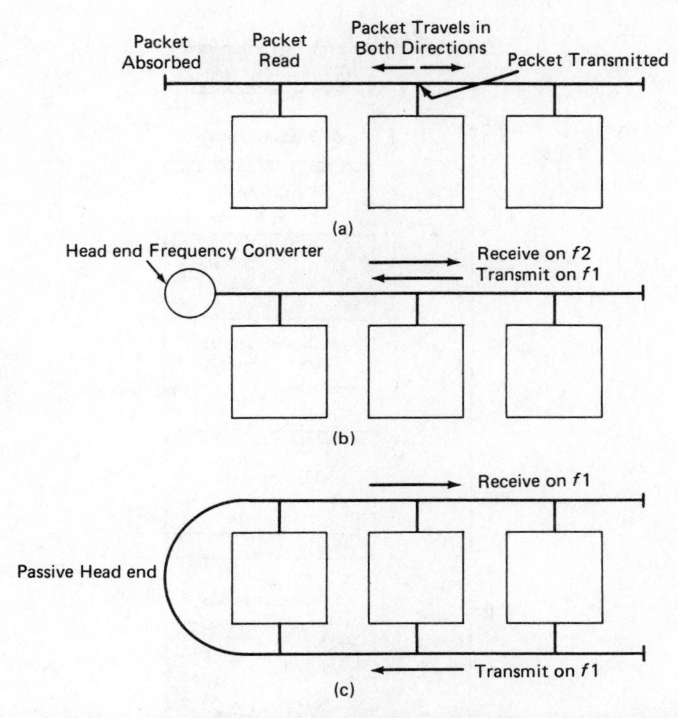

Figure 6.10. Baseband and broadband transmission: (a) bidirectional (baseband, single-channel broadband); (b) mid-split broadband; (c) dual cable broadband. (*Source*: W. Stallings, *Local Networks: An Introduction*, Macmillan, New York, 1984)

to provide point-to-point links between the repeaters of a ring. In connecting nodes to the central switch of a star topology, twisted-pair wire is commonly used.

Transmission Techniques Two transmission techniques are used in bus LANs: baseband and broadband (Figure 6-10).

A baseband LAN is by definition one which uses digital signalling. The entire frequency spectrum of the medium is used to form a digital signal which is inserted on the line as a series of constant-voltage pulses.

Baseband systems can't be extended much beyond a kilometer, because attenuation of the signal, especially at higher frequencies, causes a blurring of the pulses. Baseband transmission is bidirectional: a signal inserted into the medium at any point propagates in both directions to the ends, where it is absorbed.

A broadband LAN is by definition one that uses analog signalling. Broadband transmission is by means of radio frequency (RF) waves in a

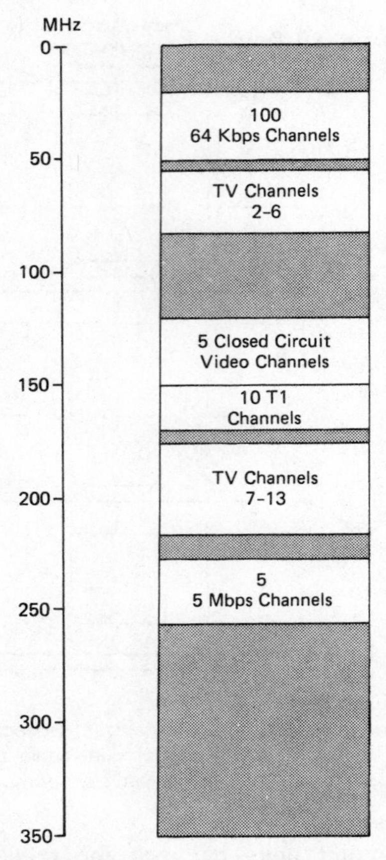

Figure 6.11. Broadband spectrum allocation. (*Source*: W. Stallings, "Local Networks," *Computing Surveys*, Vol. 16, No. 1, March 1984, p. 15)

coaxial cable medium. Analog signalling permits frequency-division multiplexing (FDM), whereby the frequency spectrum of the cable can be divided into several channels. These channels can be assigned to separate uses for video, voice, and data transmission (Figure 6-11). For data transmission, separate channels can be used to satisfy the requirements of different transfer techniques: dedicated, switched, and multiple access.

Broadband systems use off-the-shelf cable TV (CATV) components. A broadband LAN can have a diameter of well over 10 miles, and can accommodate thousands of devices. It is a unidirectional medium; complete interconnection among all nodes is achieved either by doubling the cable—one in each direction—or by a mid-split configuration, in which different sets of frequencies on the same cable are allotted to inbound and outbound paths.

A variation known as single-channel broadband dedicates the entire frequency spectrum of the cable to a single transmission path for analog signals: it therefore has the RF signal characteristics of broadband, but the same restrictions as baseband. Single-channel broadband is comparable in price and performance to baseband. A user with modest initial requirements can install a 75-ohm coaxial cable together with inexpensive modems. If he later requires a full multi-channel broadband system he can replace the modems; he will not have to go to the expense of rewiring the system.

The outcome of the extensive baseband versus broadband debate seems to be that there is room for both technologies in the LAN field. To briefly summarize the two technologies, baseband has the advantage of simplicity and low cost. The layout of a baseband cable plant is relatively simple. Baseband's potential disadvantages include limitations in capacity and distance—but these are only disadvantages if one's requirements exceed those limitations.

Broadband's strength is its tremendous capacity. Broadband can carry a wide variety of traffic on a number of channels, and with the use of active amplifiers can achieve very wide area coverage. Also, the system is based on a mature CATV technology, with reliable and readily available components. A disadvantage of broadband systems is that they are more complex than baseband to install and maintain, requiring experienced RF engineers. Also, the average propagation delay between stations for broadband is twice that for a comparable baseband system, reducing the performance and efficiency of the system.

The selection of baseband or broadband must be based on the relative cost and functional requirements. Neither can be an outright winner in the LAN wars.

6.4.2 IEEE 802 Standards for LANs

The standards issued by IEEE Technical Committee 802 contain specifications for three different types of local area networks, each conforming to a different set of needs and opinions within the industry:

1. CSMA/CD, carrier sense multiple access with collision detection, covered by IEEE 802.3 and ISO 8802/3
2. Token bus, covered by IEEE 802.4 and ISO 8802/4
3. Token ring, covered by IEEE 802.5, and ISO 8802/5

Another candidate standard, IEEE 802.6, for metropolitan LANs, is still under discussion.

It is most interesting to observe the forces shaping LAN standards. CSMA/CD is based on Ethernet, which had its start in the early 1970s at

Xerox's Palo Alto Research Center as an experiment in baseband transmission on a coaxial bus. Late in the decade Ethernet was introduced as a commercial product to interconnect Xerox office equipment, and the concept quickly caught on in the industry. In 1980 the first Ethernet technical specification, Release 1.0, was made public.

At the same time, Xerox, Intel, and Digital Equipment Corporation (DEC) announced jointly that all three would adopt the same Ethernet scheme. This announcement was issued at the time the IEEE was beginning the task of defining a standard for LAN structure and operation; its timing was an important factor in the adoption of Ethernet as the basis of IEEE standard 802.3. Slight modifications to the Ethernet specification have subsequently been made by DEC-Intel-Xerox, to bring it into closer alignment with the final IEEE standard. Today the two are virtually identical. Once the IEEE standard for CSMA/CD began to solidify, a great many Ethernet implementations began to appear.

The importance of standardization in the marketplace was strikingly illustrated in July of 1984 at the National Computer Conference in Las Vegas. A demonstration of open systems interconnection there showed that, through the use of standard protocols, equipment from different manufacturers can be made to operate in unison. In one demonstration, organized by Boeing Computer Services on behalf of the National Bureau of Standards, products from a number of different vendors were successfully tied together in an IEEE 802.3 CSMA/CD network incorporating an ISO Transport protocol. The aim, and the effect, was to prove to manufacturers that OSI standards up to the Transport layer are sufficiently well advanced to justify investment in new products embodying these standards.

Another NCC demonstration was organized by General Motors; it showed how the OSI Transport protocol could be used to link heterogeneous processors from several manufacturers on an IEEE 802.4 token bus LAN. General Motors heads a group which is concerned with the use of networks in factory automation. The group has developed a specification, the Manufacturing Automation Protocol (MAP) which incorporates the IEEE 802.4 standard, and the ISO standards for higher-layer protocols. IBM, which participated in this demonstration, is supporting the IEEE 802.4 standard with its industrial local network, and will likely base this network on MAP.*

The latest IEEE standard LAN to be specified is the token ring; work on 802.5 was completed in 1984. The existence of this standard is testimony to the clout of IBM, since it was written in large part by IBM

*By 1986 over 300 major international manufacturing companies have announced their support of MAP.

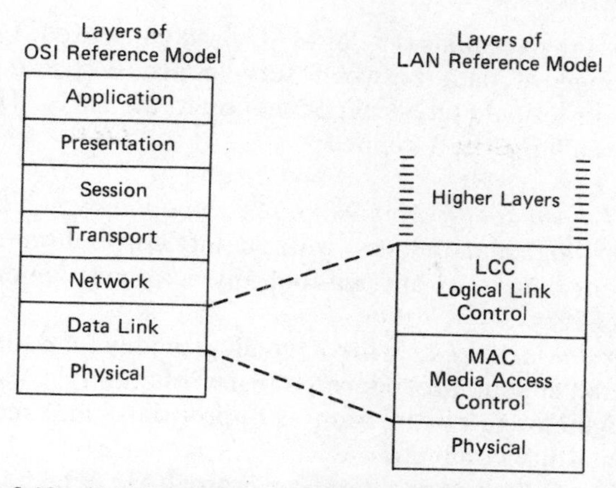

Figure 6.12. Relationship of LAN standard to OSI reference model.

representatives to the IEEE. IBM's token ring LAN product is based on IEEE standard 802.5.

IEEE 802 Standards and the ISO/OSI Reference Model The IEEE 802 standards are in the form of a three-layer architecture encompassing the functions of the Physical and Data Link layers of the OSI Reference Model (Figure 6-12).

The three IEEE 802 layers, from the top down, are Logical Link Control (LLC), Medium Access Control (MAC), and Physical. The standards offer a tree-like expansion of options (Figure 6-13).

Logical Link Control The Logical Link Control, standard 802.2, is the upper layer in the IEEE standard and is defined in a way which makes it

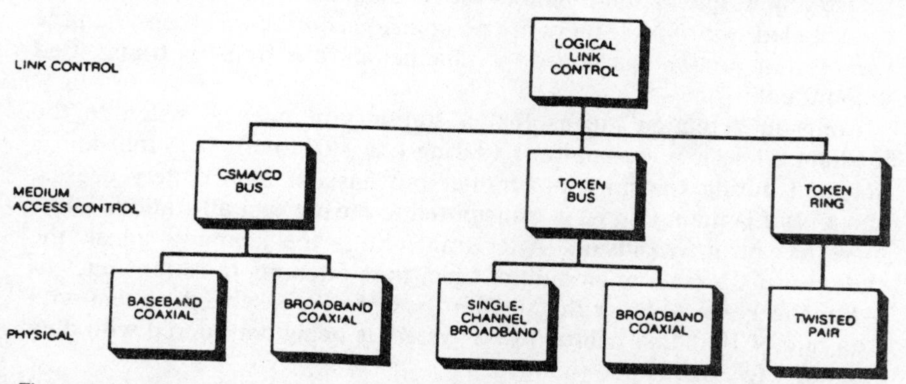

Figure 6.13. Options available in the IEEE 802 standard. (*Source*: W. Stallings, "Local Networks," *Computing Surveys*, Vol. 16, No. 1, March 1984, p. 38)

independent of the particular type of MAC procedure used. LLC provides for the exchange of data between service access points, which are multiplexed over a single physical connection to the LAN. Three types of LLC procedures have been defined:

1. *An unacknowledged-connectionless protocol* in which data units are sent over "logical data links" without any correlation to previous or subsequent data units and without any acknowledgment or guarantee of delivery
2. *A connection-oriented protocol* which provides for data link connection establishment procedure, transfer of multiple data units, acknowledgment, retransmission as appropriate, and the termination of the data link connection
3. *An acknowledged-connectionless* protocol in which a single data unit is transmitted, and then acknowledged, before a subsequent data unit is transmitted

For most LAN usage, due to low error rate and the ability to transmit large frames, connectionless services are favored.

CSMA/CD Carrier sense multiple access with collision detection, or CSMA/CD, is the first MAC protocol defined by IEEE standard 802.3 (ISO 8802/3) for the bus topology. Vendors of CSMA/CD LANs include Ungermann-Bass and Bridge Communications.

Under carrier sense multiple access a station wishing to transmit listens to the medium to determine whether another transmission is in progress. If the medium is idle, the station may transmit. Otherwise, the station continues to sense the medium until it is idle, then transmits. During transmission, the station monitors the medium for a minimum amount of time, called *slot time*. If there are no collisions during this time, then the transmitting station has seized the channel and the frame is transmitted without collision.

Collision detection means that a station continues to listen to the medium while it is transmitting (during one slot time). If a collision is detected during transmission, frame transmission immediately ceases, and a brief jamming signal is transmitted to ensure that all stations know there has been a collision. After transmitting the jamming signal, the station waits a random amount of time, then attempts to retransmit.

For the Physical layer the standard specifies a baseband system with data rate of 10 mbps. A broadband system is being considered with data rates of 1 to 5 mbps per channel.

The baseband specifications (Figure 6-14) require a 50 ohm coaxial cable. Up to 100 transceivers, or taps, may be placed on a cable segment,

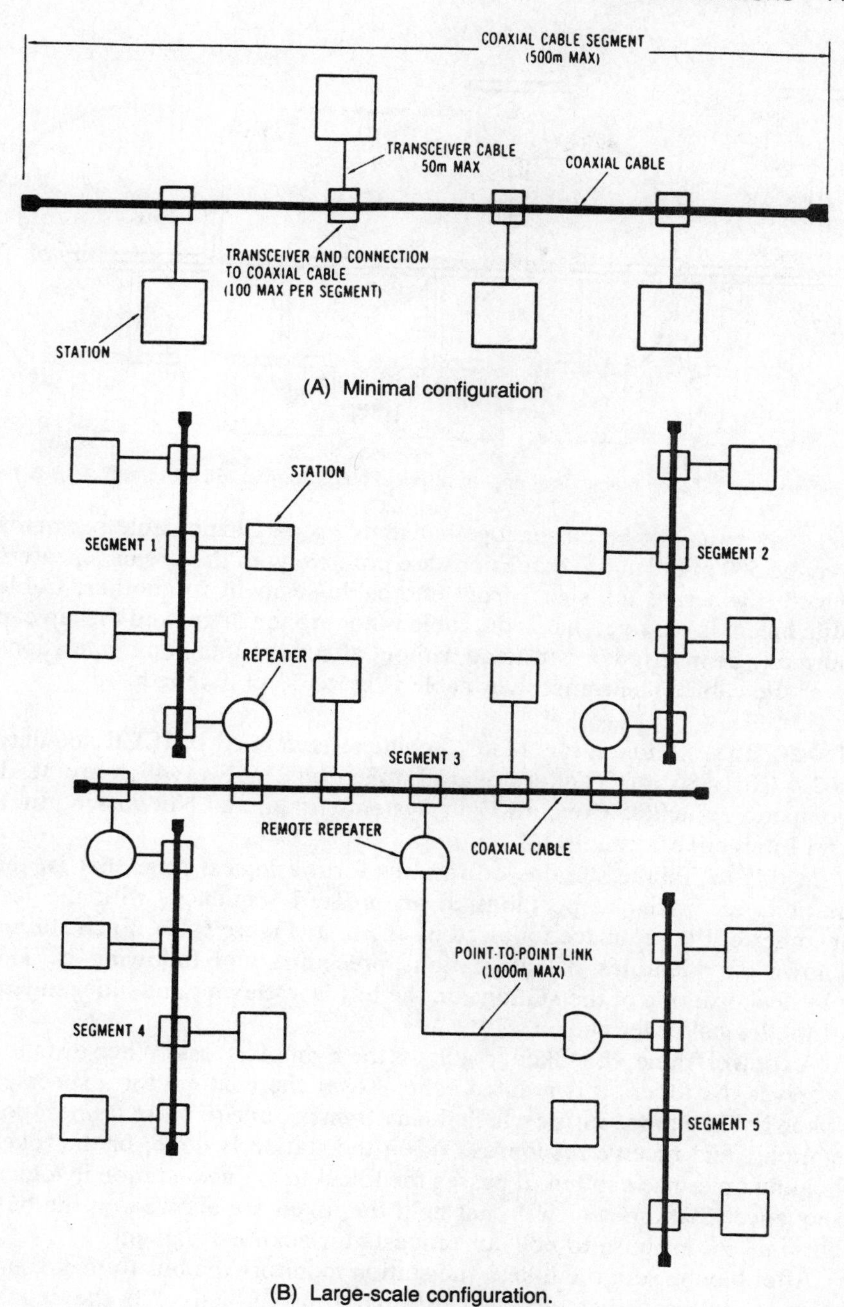

(A) Minimal configuration

(B) Large-scale configuration.

Figure 6.14. Ethernet configurations. (*Source*: Ethernet Specification.)

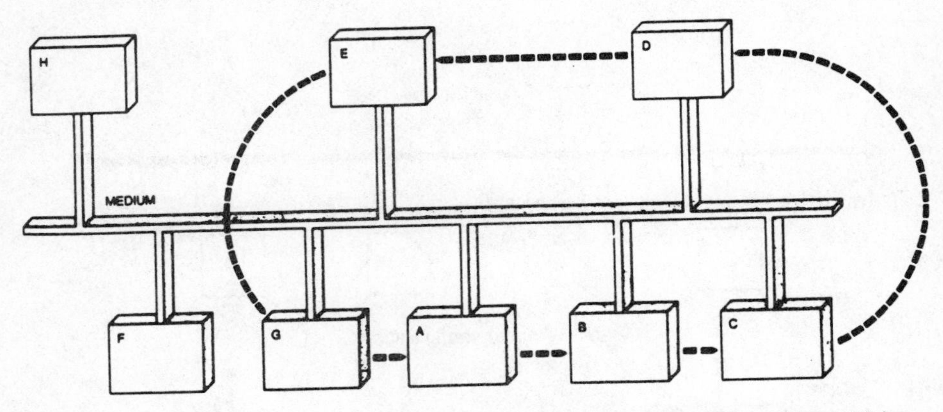

Figure 6.15. Token bus: logical ring on physical bus. (*Source*: IEEE Std. 802.4-1985.)

but they must not be closer together than 2.5 m. Single cable segments may be 500 m in length; extensions are possible with the use of repeaters which regenerate the signal from one cable segment to another. Cable attachment is passive, that is the cable is not broken or cut and the tap can normally be inserted or extracted without affecting other communications over the cable. The transceiver cable is up to 50 m in length.

Token Bus A token bus MAC technique is defined in IEEE standard 802.4 (ISO 8802/4). Vendors offering token bus LANs (which are MAP compatible) include Concord Data Systems, Industrial Networking Inc., and Interactive Systems/3M.

In this technique stations on the bus form a logical ring, that is, the stations are assigned positions in an ordered sequence, with the last member of the sequence followed by the first (Figure 6-15). Each station knows the identities of the stations preceding and following it. The physical ordering of the stations on the bus is irrelevant and independent of the logical ordering.

A control frame, the token, regulates the right of access. When a station receives the token, it is granted control over the medium for a specified token holding time, during which it may transmit one or more frames, poll stations, and receive responses. When the station is done, or the token holding timer has expired, it passes the token to the next station in logical sequence. Stations which do not hold the token are allowed on the bus, but only in response to polls or requests for acknowledgment.

After having sent the token, the station monitors the bus to make sure that its successor has received the token and is active. If the sender detects a valid frame following the token, it will assume that its successor has the token and is transmitting. Otherwise, it must assess the state of the network and, if necessary, take appropriate actions to reestablish the logical ring.

Quite a complex scheme is employed for fault management, taking care of errors such as duplicate address (two stations both think that it is their turn) or a broken ring (no station thinks that it is its turn). The standard specifies procedures for the insertion and deletion of stations in the logical ring.

Three options are provided at the Physical layer. The simplest and least expensive is a single channel broadband system using frequency shift keying (FSK) at 1 mbps. A more expensive version of this system runs at 5 or 10 mbps and is intended to be easily upgradeable to the final option, which is multichannel broadband at data rates of 1, 5, or 10 mbps.

Token Ring The last MAC protocol in the IEEE standard is token ring defined in 802.5. The token ring LAN has been receiving a lot of attention since IBM's announcement on October 15, 1985 of a token ring LAN product.

The token ring technique is based on the use of a small token frame that circulates around the ring (Figure 6-16). When all stations are idle, the token frame is labelled as a "free" token. A station wishing to transmit waits until it detects the token passing by, alters the bit pattern of the token from "free token" to "busy token," and transmits a frame immediately following the busy token. Since there is no free token on the ring, other stations wishing to transmit must wait. The frame on the ring will make a round trip and be purged by the transmitting station.

A token holding timer controls the length of time a station may occupy the medium before passing the token. The transmitting station will insert

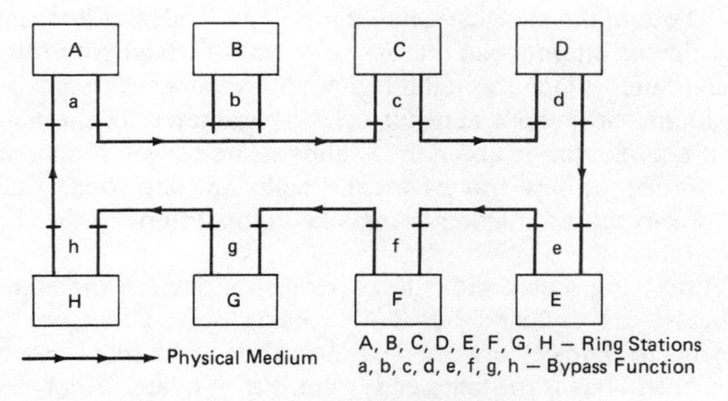

Figure 6.16. Token ring configuration. (*Source:* IEEE std. 802.5-1985, p. 24.)

a new free token on the ring when both of the following conditions have been met: the station has completed transmission of its frame, and the busy token has returned to the station. When a new free token is released, the next station downstream with data to send will be able to seize the token and transmit.

For the Physical layer twisted pair wiring has been defined, with data rates of 1 and 4 mbps.

6.4.3 Comparing LANs

Table 6-4 is a comparison of the attributes of the three different standard medium access methods. To illustrate the performance of the standard LAN access methods, Figure 6-17 depicts the results of a study carried out by a group at Bell Laboratories [Stallings 1984]. The study compares network throughput under extreme load conditions and with different frame sizes. Given 100 stations, heavy load conditions assumes that all 100 stations have data ready for transmission. Under light load only one station has data ready for transmission. In both cases the input is enough to fully utilize the network. The length of the medium is assumed to be two kilometers. The token ring has 1 bit latency per station. Two frame sizes are used: 500 and 2000 bits per frame.

The graphs depict maximum potential utilization given the actual capacity of the medium. The results show that token ring is the least sensitive to load conditions, whereas CSMA/CD is the most sensitive. Under the given conditions, a smaller frame size causes a greater difference in throughput between token passing and CSMA/CD. The reason is that smaller frames mean more frames, which results in higher collision rate under CSMA/CD. Token bus requires large delays even under light load conditions due to token processing time in each station. CSMA/CD offers the shortest delay under light load, but it is unstable in the sense that as offered load increases, so does throughput until, beyond some maximum value, throughput actually declines as load increases. The maximum throughput depends on the parameters of the network (in graph b the maximum is about 1.25 mbps). The reason is that with high collision frequency few frames escape collision, and those that collide must be retransmitted, further increasing competition for the medium.

Bus vs Ring For a user with a large number of devices and high capacity requirements, a broadband bus LAN seems best. For more moderate requirements the choice between a baseband bus and a ring is not clear cut.

The baseband bus is the simpler system. Passive taps rather than active repeaters are used. Thus, medium failure is less likely, and there is no need for the complexity of bridges and ring wiring concentrators.

Table 6.4. Comparison of Standard Medium Access Methods.

	CSMA/CD		TOKEN BUS			TOKEN RING
Physical medium	50 ohm baseband coaxial	75 ohm broadband coaxial (under consideration)	75 ohm single channel coaxial pair		75 ohm broadband coaxial	150 ohm baseband twisted
Signalling technique	Manchester	to be defined	phase continuous FSK	phase coherent FSK	multilevel duobinary AM/PSK	differential Manchester
Data rate (mbps)	10	1–5 per channel	1	5 or 10	1, 5 or 10	1 or 4
Cost	low	high	low	medium	high	high
Address size (bits)	16 or 48		16 or 48			16 or 48
Network size (meters)	2800 maximum diameter		7600, 1 mbps based on attenuation for other frequencies			not specified
Maximum stations	1024		not specified			not specified
Minimum packet (bytes)	56 or 64		9 or 17			3
Maximum packet (bytes)	1518		8191			not specified
Maturity	yes		yes			no
Security	low		medium			high
Error rate	medium		medium			low
Complexity	low		high			high
Response time at light load	fast		slow			fast
Response time at heavy load	slow		medium			fast
Maximum transmission delay	no		yes			yes

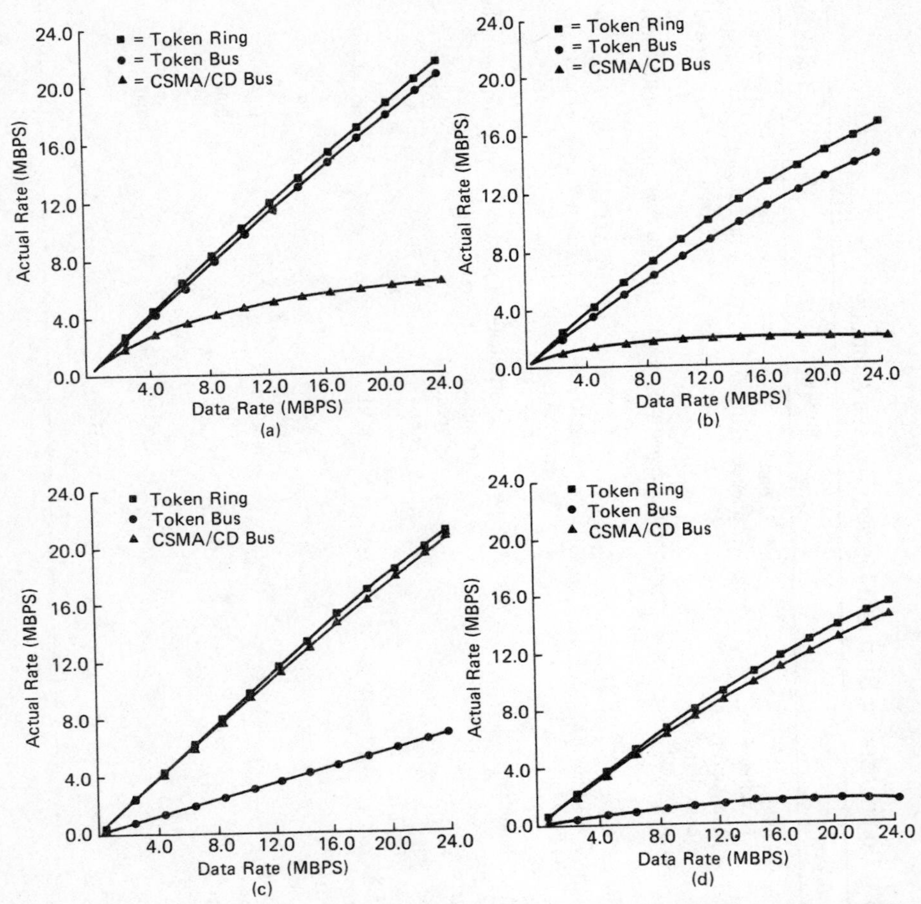

Figure 6.17. Potential throughput of IEEE 802 protocols: (a) 2000 bits per frame, 100 stations active out of 100 stations total; (b) 500 bits per frame, 100 stations active out of 100 stations total; (c) 2000 bits per frame, 1 station active out of 100 stations total; (d) 500 bits per frame, 1 station active out of 100 stations total. (*Source*: W. Stallings, "Local Networks," *Computing Surveys,* Vol. 16, No. 1, March 1984, pp. 34–35.)

The most important benefit of the ring is that, unlike the bus, it uses point-to-point communication links, which has a number of implications. Transmission errors are minimized and greater distances can be covered since the transmitted signal is regenerated at each node. The ring can accommodate optical fiber links, providing very high data rates and excellent electromagnetic interference characteristics. The electronics and maintenance of point-to-point lines are simpler than multipoint lines. Another benefit of the ring is that fault isolation and recovery is simpler than for bus. On the other hand, tuning and maintenance of the LAN are more difficult in a ring network.

The ring as well as the token bus provide a guaranteed maximum transmission time, an important characteristic for applications such as process control.

CSMA/CD vs Token Bus The principal disadvantage of token bus is its complexity; the logic at each station far exceeds that required for CSMA/CD. A second disadvantage is the overhead involved; under light load conditions a station may have to wait through many fruitless token passes for a turn.

Advantages of token bus are first that it is easy to regulate the traffic; different stations can be allowed to hold the token for different amounts of time. Second, unlike CSMA/CD there is no minimum frame length requirement for token bus. Third, the requirement for listening while talking imposes physical and electrical constaints on the CSMA/CD system that do not apply to token systems.

Token bus is significantly superior to CSMA/CD under heavy load. In fact there is a known upper bound to the amount of time any station must wait before transmitting. The upper bound is known because each station in the logical ring can hold the token for a specified time. In contrast, the delay time with CSMA/CD can only be expressed statistically; moreover, there is a possibility that a station could be shut out indefinitely, if every attempt to transmit ends in a collision.

For process control and other real-time applications the nondeterministic behavior of CSMA/CD is undesirable. In reality, a token bus system has also a statistical component to its behavior, since there is a possibility for a transmission error which can cause a lost token.

For performance studies of LAN protocols see Bux [1984], Stuck and Arthurs, [1985], and Stallings [1984].

6.5 LONG-HAUL COMMUNICATIONS

In the context of open systems, this title refers in general to the means of interlinking geographically separate systems over distances ranging from tens to thousands of miles. As such they complement LANs in the spectrum of open system communications.

6.5.1 Introduction

Early work on local area networks was largely based on knowledge gained in long-haul communications. For example, the work which led to

Ethernet was initially an extension of the Aloha ground radio packet broadcasting network at the University of Hawaii.

Much of our present knowledge of networking is derived directly from the ARPANET long-haul network project, a creation of ARPA (now DARPA), the (Defense) Advanced Research Projects Agency of the U.S. Department of Defence. From the late 1960s onward DARPA has stimulated research in computer networking by providing grants to computer science departments at many American universities, as well as to a few private corporations. This research led to an experimental four-node network which became active in December of 1969, and has subsequently grown to well over 100 computer sites spanning half the globe, from Hawaii to Norway.

The general characteristics of long-haul communications are:

- Long distances, often tens of thousands of kilometers.
- The use of public carriers which own the communications facilities. The main contributors to the development of long-haul computer communications have been the communications companies, not the computer industry.
- Low-bandwidth communications channels.

Table 6-5 compares the characteristics of long-haul networks and local area networks.

The main choices in long-haul communications are between analog and digital services, between terrestrial and satellite services, between dedicated and switched services, and (in the U.S.A. since deregulation of the communications industry) among public carriers.

Analog or Digital Service Analog channels are voice-oriented; they were originally intended for voice communications. Digital channels are data-oriented. The lower cost and better transmission quality of digital transmission are desirable for voice communications as well; a gradual transition from analog to digital is therefore under way in North America.

The telephone industry is planning to convert 90% of local and 40% of long-distance interoffice transmission links to digital system by 1990. It is anticipated that within 10 to 20 years analog lines will be obsolete—although currently the great majority of communications links are analog.

Availability is a key issue. Analog lines are available virtually everywhere, while digital service is still rather sparse. AT&T's Dataphone Digital Service (DDS) is currently the largest digital service, serving about 100 cities. Digital facilities, however, will become more widely available, and it is generally agreed that they will provide better service, mainly because of fewer line-related transmission errors and a greater "uptime."

Table 6.5. Comparison of Local and Long-Haul Network Characteristics.

CHARACTERISTIC	LOCAL AREA NETWORK	LONG-HAUL NETWORK
Typical Bandwidth	10 million bits per second.	56,000 bits per second.
Acknowledgment	One message acknowledged at a time.	*N* messages acknowledged at a time.
Message Size and Format	Small (simple header). No need to divide message into packets.	Large (complex header). Need to divide message into packets.
Network Control	Minimum requirement due to small number of links and nodes and simple topology	Extensive due to large number of nodes and links and complex topology.
Flow/Congestion Control	Minimum due to high bandwidth and simple topology.	Extensive due to low bandwidth and complex topology.
Error Rate	Relatively low. Operated in benign environment.	Relatively high. Operated in noisy environment of telephone network.
Message Sequence and Delivery	Minimum problem due to simple topology (e.g., bus or ring).	Major problem due to complex topology (e.g., mesh).
Standard Architecture	Usually only two or three bottom layers provided.	Frequent use of all or many ISO layers.
Routing	None required due to simple topology.	Major problem due to complex topology.
Delay Time	Small due to short distance and medium (e.g., coaxial cable).	Large due to distance and medium (e.g., satellite).
Addressing	Simple intra-network communication due to simple topology. Complex inter-network communication due to the use of long-distance network(s).	Complex because of many nodes and links.

SOURCE: N. F. Schneedewind, "Interconnecting Local Networks to Long-Distance Networks," *Computer*, Sept. 1983, IEEE, pp. 15–24.

Another important issue is speed. The most common type of analog line is the voice-grade channel (3002 in AT&T's designation). Such channels support speeds as high as 16 kbps with appropriate modems. If higher speeds are required, the next increment is the wideband channel (8000 series). These cost about 7 to 10 times more than voice-grade channels, and support speeds up to 56 kbps. DDS supports the following speeds: 2.4 kbps, 4.8 kbps, 9.6 kbps, 56 kbps, and 1.544 mbps.

A final issue is the alternate use of the lines for voice and data. Current equipment makes it cost-effective to alternate between data and voice on a shared analog channel. Use of a voice-grade line for data transmission during off hours can provide significant savings. The same capability is not commonly available for digital lines, although with the widespread of ISDN the situation will change.

Terrestrial or Satellite Although satellite channels may cost less and offer lower error-rates, they do so at the cost of much higher propagation delays. Because satellites orbit the earth at an altitude of 35,800 kilometers (the Clarke orbit), there is a propagation delay of about ¼ second; the effects are a low throughput and poor response-times as compared with those of terrestrial links.

Switched or Dedicated Another consideration in circuit selection is the choice of dedicated or switched services. For a fixed monthly cost, dedicated lines (also called leased or private lines) provide a permanent connection requiring no dialing or setup procedures. On the other hand, switched or dial-up services require call setup and are charged on a usage or time-sensitive basis. They provide flexibility by allowing the user alternate access to multiple points, through the call setup procedure. Dial-up lines operate at low speeds, and are relatively inexpensive. The user has no control over the quality of the circuit. On a leased line, by contrast, the circuit is often conditioned for high-speed, high-quality transmission.

6.5.2 Long-Haul Technology

This section contains a discussion of some of the more significant aspects of long-haul communications technology.

High-Speed Dial-Up Modems Modems make it possible to transmit computer data over ordinary telephone lines by modulating digital signals at the transmitting end and demodulating the received signal back into its original form at the receiving end (Figure 6-18).

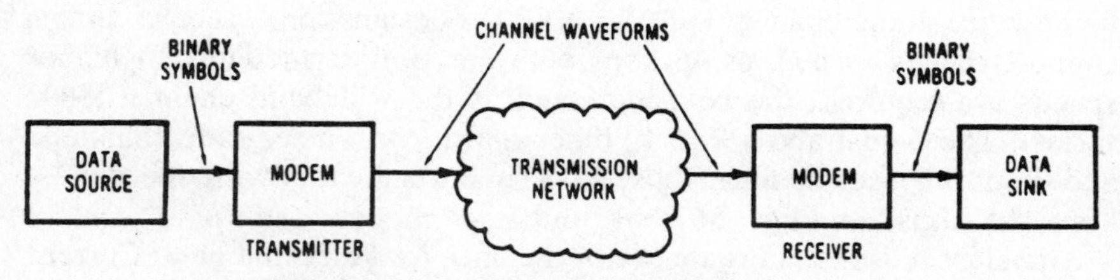

Figure 6.18. A one-way point-to-point data link.

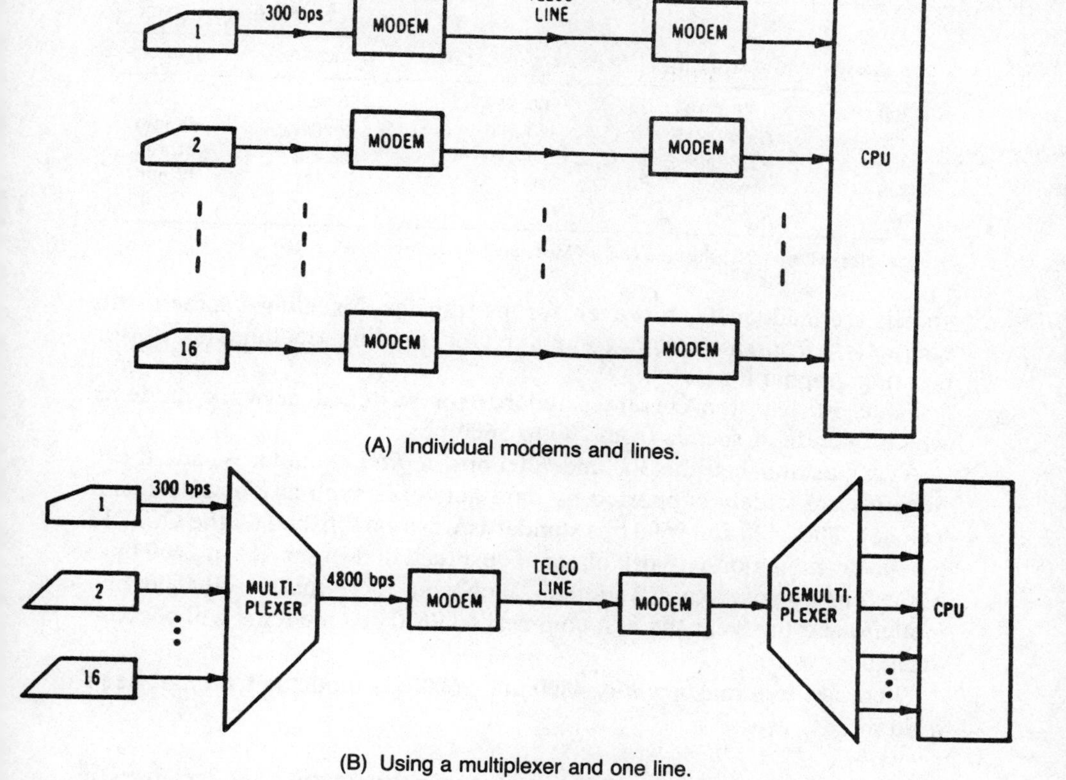

(A) Individual modems and lines.

(B) Using a multiplexer and one line.

Figure 6.19. Using multiplexers to save costs. (*Source*: R. L. Pickholtz, "Modems, Multiplexers, and Concentrators," *Data Communications, Networks, and Systems*, Chapter 3 (T. C. Bartee, Editor-in-Chief), Howard W. Sams & Co., 1985, pp. 64–66.)

One way of reducing the cost of connecting multiple terminals to a computer via modems is by using a multiplexer device (Figure 6-19). The multiplexer combines the traffic on a number of connected terminals, prior to their transmission via a high rate modem. At the receiving end a demultiplexer separates the transmission to its original form.

At least 70% of all modems are used on the switched telephone network, and the rest on dedicated lines. Two types are dominant among the switched network modems: AT&T 103 for speeds up to 300 bps, and AT&T 212 for speeds up to 1200 bps.

By way of comparison, a fast typist (100 words a minute) produces data at about 100 bps; the average person can read text as it appears on a screen at 300 bps. Standard modems handle such applications well. Their

Table 6.6. Comparison of standard Dial-up Modems

SPEED (BPS)	STANDARD	COST RANGE ($)	TRANSMISSION TIME	
			TO SCREEN	TO FLOPPY
300	AT&T 103	60–350	66 sec.	3.2 hours
1200	AT&T 212	450–700	17 sec.	48 min.
2400	V.22 bis	800–1400	8.3 sec.	24 min.
4800	none	2500	4.2 sec.	12 min.
9600	V.32	n/a	2.1 sec.	6 min.

SOURCE: Maxwell Kim, "High-speed Dial-up Modems," Byte, Dec., 1984, p. 180.

speeds are inadequate, however, for file transfer, for filling a screen with characters from a remote computer, for sending documents, or for creating graphic images.

Table 6-6 lists the current standards for switched network modems which operate at speeds from 300 to 9600 bps.

AT&T established the 300 and 1200 bps *de facto* standards, and these modems are widely supported by data networks such as TYMNET and Telenet. The 2400 and 9600 bps standards were established by the CCITT. However, no major network offers widespread dial-in service at 2400 bps. Some companies, e.g., Anderson Jacobson Inc., market a 4800 bps modem, and this year the first commercial 9600 bps modems will become available.

There are two reasons why 4800 and 9600 bps modems have not been used in past years:

1. The end connection of the switched network to a modem uses a single two-wire line, carrying the signal in both directions. High-speed modems were designed for leased lines, which normally carry two two-wire lines to the modem, creating separate signal paths for sending and receiving data.

2. Modems used on the switched telephone network must be full-duplex: that is, capable of sending and receiving data in both directions simultaneously. The reason for this requirement is that in the early days terminals were connected directly to computers with separate paths for data sent and received. Communications interfaces and software therefore assumed a full-duplex connection. Bell's first switched-network modem, which appeared in the early 1960s, provided full-duplex operation by dividing the telephone line into two subchannels. The cost of converting all networks, computers, and terminals now installed to half-duplex would be prohibitive. Switched-network modems therefore

generally provide full-duplex data transmission over two-wire circuits with a bandwidth (frequency range) of 300 to 3400 Hz.

A complicated set of techniques is used to squeeze full-duplex data transmissions at 9600 bps onto a bandwidth of 3100 Hz. This includes phased modulation, fitting and echo cancellation. For details see Kim Maxwell [Maxwell 1984].

Implementing 300 and 1200 bps modems on single chips has kept prices within affordable bounds. In 1986 2400 bps modems based on two or three chips entered the market. Custom VLSI is one factor that will reduce the price, and make possible the general use of higher-speed modems over the next few years. Other factors are volume production and an increased market requirements for such devices.

Satellite Communications A communications satellite is essentially a big repeater in the sky: it simply amplifies incoming signals and retransmits them with a transponder. (A transponder is a device which receives radio-frequency signals at one frequency and converts them for retransmission at another frequency.) Its big advantage over an earth station is that it offers complete interconnection among all the stations using it.

In order to connect n stations fully, a satellite system must provide n carriers, each one using a different frequency in both the upward and downward direction. To interconnect n stations with terrestrial lines requires $n(n - 1)/2$ *links*.

A communications satellite like INTELSAT VI has 48 transponders, each with a capacity of 40, 80, or 160 MHz; it can support 30,000 simultaneous telephone conversations. Communications satellites are put into geosynchronous orbit to make them appear stationary when viewed from the earth. A ground-station antenna, once installed, need not be moved; it is therefore a relatively cheap device.

A non-geostationary satellite would require a much more expensive steerable antenna, and might sometimes have the added disadvantage of being on the wrong side of the earth. On the other hand, the high altitude required to achieve a stationary appearance entails an up-and-down signal propagation delay of 270 milliseconds.

Three modes of interaction with the end user are common. The most direct, but more expensive one, is to install a complete ground station on the user's roof. The cost of such stations is declining rapidly. The Equatorial Communications Company, for example, is producing a micro earth station antenna with a 24-inch diameter at a cost of less than $3000. Many such earth stations are used as input devices for personal computers in distributed applications. Low-cost interactive micro earth stations are being developed for connecting both personal computers and terminals to remote hosts.

The second mode of interaction is to place a small, cheap antenna on the user's roof to communicate with a shared satellite ground station on a nearby hill. The third approach is to have the user gain access to the shared ground station via a leased telephone circuit or a television cable.

In addition to providing wide transmission bandwidths, satellite channels provide low error-rates. Signals from terrestrial stations run up and out of the earth's atmosphere directly to the satellite, and proceed down from the satellite to the receiving station. Of the many miles the signal travels, only about ten miles are through the earth's atmosphere. By comparison, terrestrial microwaves may pass through the dense lower atmosphere, more or less parallel to the ground, for thousands of miles; they are therefore much more subject to atmospheric interference. Satellite channels have typical error-rates of 1 in 10^9 bits transmitted; for typical analog terrestrial circuits the figure is 1 in 10^5 bits. The error-rates offered by digital terrestrial channels (e.g., AT&T's Digital Dataphone Service) are equivalent to those of satellite channels.

In the future, on-board demodulation of uplink time-division multiple access (TDMA) and frequency-division multiple access (FDMA) carriers will be combined with switching at the individual channel level. Remodulation of channels into downlinks will be by means of both fixed and hopping narrow beams, to permit optimum use of the satellite for carrying both high-volume telephone trunks and various forms of thinner traffic. Such a system will require sophisticated network route control facilities to send sufficient routing instructions to the on-board switch and to synchronize the traffic bursts.

Challenges to current satellite technology will come from several directions:

Fiber optics. Laser transmission at 1 GHz (a billion cycles per second) will be highly competitive for simple point to point trunk connections over distances greater than a thousand kilometers. Impressive recent progress in single-mode fiber technology gives the promise of 4 gbps transmission at relatively low cost.

Terrestrial microwave services will soon be competitive for transmission paths of less than 100 kilometers.

Trade and political pressures will encourage proliferation of conventional satellites directed at national markets, rather than the development of technically more advanced satellites to consolidate multinational coverage. Restrictions on technology transfer will tend to retard realization of the potential economies of scale offered by satellites. In the current global environment, there is a strong trend toward proliferation of smaller, less innovative satellites, and to increasing saturation of the geosynchronous Clarke orbit.

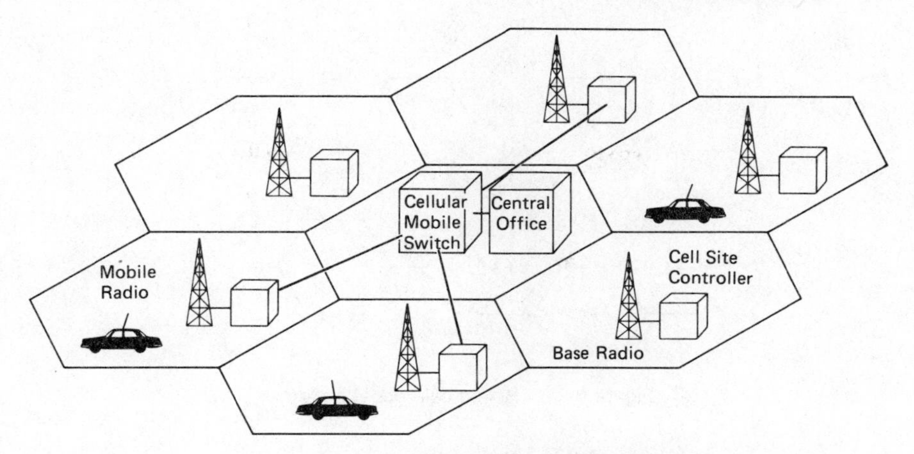

Figure 6.20. Cellular mobile radio system.

Cellular Radio Telephone Cellular radio is a means of providing mobile telephone service using radio frequency transmission. The term was first used to describe a proposal by Bell Laboratories in 1947 for an advanced form of mobile radio. In a traditional mobile telephone system a radio tower is placed at the center of a city, and serves mobile units within a radius of 25 to 70 miles. In most areas the number of channels available is between 12 and 40.

With so few channels it is possible to serve only a limited number of customers. Moreover, during peak hours there is a long waiting period for service. The radio signal between the mobile phone and the central tower is often of poor quality because of fading, static, and interference from other systems. Moreover, the strong signal emitted from the radio tower prevents the use of the same frequencies by radio towers in nearby cities (because of interference).

In cellular technology a city is divided into many small areas, or cells, and a radio tower is placed at the center of each cell (Figure 6-20). Each of these towers has a low-power transmitter, strong enough to reach the edge of the cell but not much beyond. The mobile radiophone emits signals which are picked up by the tower in the cell in which the phone is currently located (i.e., through which it is passing).

As the telephone (normally in a motor vehicle) moves toward the edge of a cell, its signal begins to fade in that cell, and at the same time to grow stronger in the adjacent cell, into which it is moving. A central switching office constantly monitors the mobile radio's signal strength in cell, and at some point it decides to hand off the call from one cell to the next.

Because all transmissions are at a low power, the same frequencies can be used in many cells, so long as these cells are not immediately adjacent

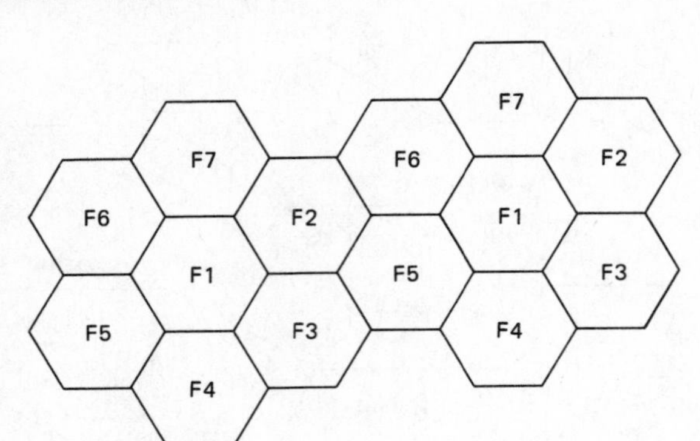

Figure 6.21. Seven cell repeat pattern.

(see, for example, Figure 6-21). Since the same frequencies can be reused many times through the system, the capacity of the telephone system is greatly increased. Moreover, if one cell becomes congested it can be subdivided into smaller cells.

Customers use cellular radio as if it were a normal telephone. It can also be used for data communication, currently at 300 bps; in the near future data communication lines will be offered, allowing higher data rates.

Digital transmission is a way of making more efficient use of the frequency spectrum. Futuristic personal communications devices will probably transmit digitally and operate through cellular land-based networks as well as direct satellite networks. Such devices will ultimately combine telephone, data transmission, broadcast radio, TV, as well as personal computing facilities [Crump 1985].

6.5.3 Connections and Protocols

In this section a number of specific protocols are described in the context of the ISO open systems interconnection model. The lower layers of the model are treated first, since the historical progression of protocol development has been from the bottom up. Protocols at the Physical and Data Link layers came first; developments at the Transport layer and above have been quite recent. The description of the layers is followed by a discussion of the connection and connectionless modes of data transmission.

The Physical Layer The function of this layer is to permit the host or terminal to send or receive a ''raw'' bit stream into or from the network. The best known Physical layer standard is RS-232-C, which specifies the meaning of each of the 25 pins on a terminal connector, as well as the protocol governing the use of each. A newer standard, RS-449, is "upward compatible" with RS-232-C, and uses a 37-pin connector to accommodate a greater number of signals.

The CCITT has developed the X.21 standard, which specifies the interface between a host computer and a network using a 15-pin connector with more logic. X.21 is a circuit-switching protocol, providing a connection which remains open so long as the host wishes to communicate with the network. In X.21 terminology, the connection is established between DTE (data termination equipment) and DCE (data circuit terminating equipment), the latter being a network switching node. The host-to-host connections set up over an X.21 line may be either circuit- or packet-switched.

The Data-Link Layer The responsibility of the Data Link layer is to detect and correct transmission errors, and to control the flow of data between a fast sender and a slow receiver. The most important protocol at this level is the ISO's High-level Data-Link Control (HDLC); a number of other Data Link protocols, such as IBM's SDLC, differ from HDLC in minor ways.

The HDLC frame format is illustrated in Figure 6-22. HDLC uses a 01111110 pattern to delimit frames. A "bit-stuffing" technique is used to distinguish between this pattern as a delimiter and its occurrence as part of a message. That is, whenever five consecutive 1-bits appear in the data stream, a 0-bit is stuffed into the stream (usually by hardware). Six consecutive 1-bits therefore cannot appear except in the frame delimiting sequence.

An address-field is used for addressing multipoint lines. The control-field is different for each of three classes of frame: information, supervi-

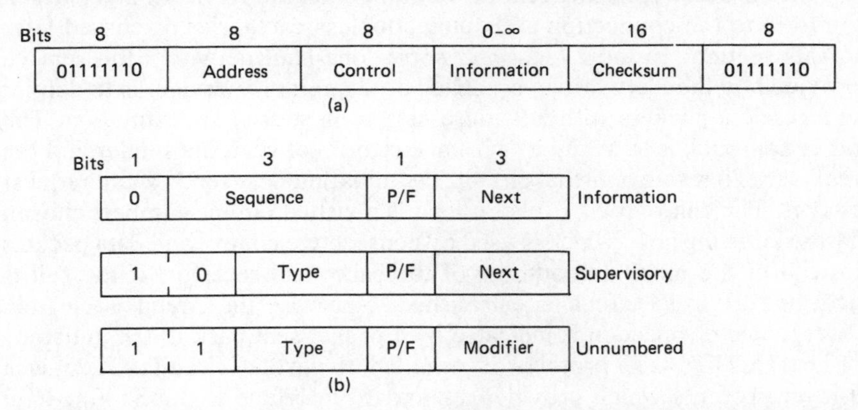

Figure 6.22. (a) The HDLC frame format. (b) The control byte for the three kinds of frames.

sory and unnumbered. For example, in information frames (i.e., those which contain ordinary data) the sequence- and next-fields contain the sequence number of the current frame and of the next frame expected.

The procedure of attaching an acknowledgment to an outgoing data frame (next) is known as "piggybacking." This practice saves bandwidth because it requires fewer frames, and therefore reduces overhead on the receiving machine (fewer I/O interrupts).

When there is no reverse traffic on which an acknowledgment can be piggybacked, a separate supervisory frame is used. Other types of supervisory frames are for negative acknowledgment, selective repetition, and indication that the receiver is temporarily not ready. The P/F (poll/final) bit has various uses, such as identifying polling frames on multipoint lines, and identifying the final frame in a sequence.

Unnumbered frames are used for miscellaneous control information, such as line initialization and the reporting of abnormal conditions. The checksum is a polynomial code (also known as a cyclic redundancy code) used in error detection. If errors are detected a request for retransmission will follow.

The Network Layer The primary task of the Network layer is to perform routing of protocol data units from source to destination nodes via, if necessary, intermediate switching nodes. When a frame arrives at the network switching node, the Data Link layer strips the frame header and trailer; what is left, called a packet, is transferred to the Network layer. This layer then chooses an outgoing line on which to forward the packet. In performing this routing function it tries to balance the traffic on overloaded and underloaded lines, thus exercising a congestion control function as well.

There are two opposing schools of thought on the Network layer which conform to the connection and connectionless categories discussed later in this section. In most local and some long-haul networks the service provided by the Network layer, called a "datagram" service, is to deliver independent packets with full addressing from source to destination. The other approach is to set up a "virtual circuit" between the sender and the receiver. To set up a virtual circuit, the initiating host sends a call request packet. The call request packet contains a virtual circuit number, chosen by the initiating host, which is used by consecutive control and data packets instead of the name and address of the caller and receiver. If the call is accepted a virtual circuit is established; otherwise the circuit is cleared. Acceptance or rejection is indicated by a packet sent back to the initiator.

The CCITT X.25 protocol is oriented to virtual circuits. X.25 is a 3-layer protocol which provides a standard interface to public long-haul networks (Figure 6-23). The Physical layer is defined by X.21 (or by X.21

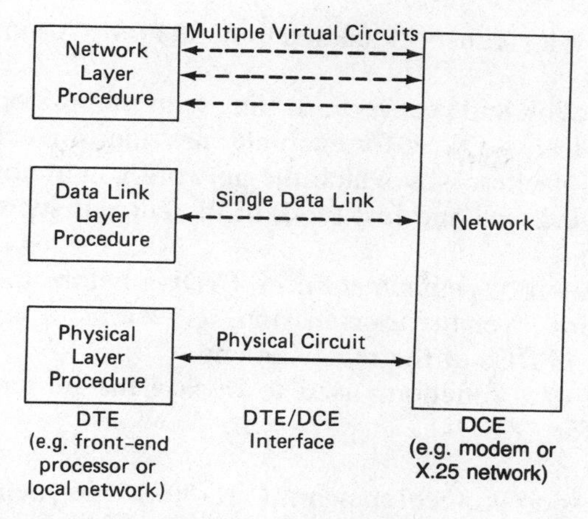

Figure 6.23. Structure of the X.25 interface (*Source*: N. F. Schneidewind, "Interconnecting Local Networks to Long-Distance Networks," *Computer*, Sept. 1983, pp. 15–24.)

bis, which defines an interim analog interface). The Data Link layer consists of two HDLC variants, link access procedure (LAP) and link access procedure, balanced (LAPB).

Under X.25 there is some overlap between responsibilities of the Network and Data Link layers, for example in providing next and sequence numbers for flow control and sequencing. However, Data Link layer sequence-number and acknowledgment refer to traffic between host and DCE for all virtual circuits combined. Data Link flow control is required to prevent the host from flooding the DCE. By contrast, in the Network layer the sequence numbers are for individual virtual circuits and control individually the flow of each such circuit. The Network layer also uses control packets to reset a particular virtual circuit, to reset all virtual circuits following the crash of a host or DCE, to perform acknowledgments or interrupts, and to indicate temporary problems.

The Transport Layer The Transport layer is a logical separator between the network proper and the Session layer. It makes it possible to plug in different low-level protocols without changing the Session layer or layers above it—just as a compiler hides true machine intructions from the user of a high-level language. Thus the Transport layer is required to provide reliable end-to-end communication.

Major functions within the transport layer are:

1. Transmission of transport protocol data units (TPDUs)
2. Multiplexing of several Transport connections on a single Network connection

3. Splitting of a Transport connection into two or more Network connections
4. Error detection and recovery, dealing with the loss or reordering of packets in transit (X.25, for example, provides a mechanism—reset and restart packets—by which the network can announce to a host that it has crashed and lost track of the current sequence numbers for packets)
5. Concatenation or segmentation of TPDUs before their delivery to the Network layer for transmission, as well as separation or reassembly of TPDUs at the receiving end
6. Expedited data functions used to bypass the normal flow control provided for TPDUs

The ISO Transport protocol standard, [ISO 8073 1986], defines five classes of protocol. A class is a set of functions, some of which may be optional. The choice of classes and options is negotiated during the establishment of a connection, on the basis of user requirements, the quality of available network services, and the cost of these services. To provide the basis for a decision on the class of Transport protocol to be used on top of a given Network connection, Network services are classified in terms of their error behavior. Table 6-7 summarizes the types of networks and the Transport classes recommended for use on top of each.

Acceptance of the ISO Transport protocol standard will be accelerated by two recent developments:

1. The decision by the ten member-governments of the European Community to adopt procurement policies which will require OSI standard protocols for all government-related purchases.
2. The publication by the U.S. National Research Council's Commit-

Table 6.7. Mapping Transport Classes to Network Types.

NETWORK TYPE	A		B		C
Residual error rate	acceptable		acceptable		unacceptable
Signalled error rate	acceptable		unacceptable		unacceptable
Transport class	0	2	1	3	4
Transport functions	simple class	multiplexing	error-recovery	error-recovery, multiplexing	error-detection, error-recovery, multiplexing

tee on Computer/Computer Communication Protocols of a report on "Transport Protocols for Department of Defence Data Networks." This report recommends the adoption of the ISO Transport and Internetwork protocols as alternatives to and eventual replacements for the current Defence Department Transport (TCP) and Internet (IP) protocols. The recommendation was adopted by the U.S. Department of Defence.

The Session Layer The primary task of the Session layer is to connect two processes together to form a connection called a session. By mapping them onto Transport addresses, the Session layer allows users to refer to destinations by symbolic names. When a session is set up, an activity called "session binding" establishes the conventions which govern the session: full- or half-duplex data transfer, negotiated session release, expedited data, etc.

The dialog control task performed by the Session layer is useful in networks in which user-primitives for sending and receiving messages are nonblocking, and where multiple requests may be outstanding on a single session at the same instant. The dialog control function keeps track of requests and replies, reordering them to simplify the design of user programs.

Another aspect of dialog control is the bracketing of groups of messages into atomic units. In many database applications it is undesirable that any transaction be broken off part way through, as a result of network failure, for example. If the transaction pertains to a group of messages, the Session layer can ensure that the entire group has been received at the destination before permitting the transaction to begin.

The ISO Session protocol standard is [ISO 8327 1984].

The Presentation Layer The Presentation layer is concerned with the syntax of the information communicated between applications. The semantics of the information are known and relevant only to the applications themselves. The Presentation layer service allows systems to communicate about the syntax of Application layer information exchanges, but not about syntax within the systems themselves. Where there are differences between syntax in a local systems and the transfer syntax, mapping between the two must be carried out in the Presentation layer.

The ISO service definition and protocol specification for the Presentation layer [ISO 8822, 8823 1986] are currently at the draft international standard stage.

The Application Layer The ISO is still in the process of defining the Application layer. As specified in the OSI Reference Model, Application "service elements" call upon one another, as well as on Presentation services, to provide communications to a "user element." The currently emerging standard defines two types of application service element: common and specific, or CASE and SASE. The distinction between these two is somewhat arbitrary, but is based on the notion that CASEs provide services which are used in most applications, while SASEs are specific to particular applications. Some SASEs which are of broad utility will be standardized, and become CASEs.

Work on CASEs has been devoted thus far to providing a basic kernel subset, plus a service ensuring commitment, concurrency, and control [ISO 8649, 8650 1986].

In SASEs, the following services have been defined:

1. *Virtual terminal,* which permits users to communicate in the form of a dialog while supporting the transfer and manipulation of graphic image data by both users.
2. *File transfer, access and management,* in which users of an open system can transfer, have access to, and manage files in any other remote open system. Since the implementations of file storage vary considerably, a common model for describing files and their attributes is established, and mapped to particular local systems [ISO 8571 1986].

Connection-Mode vs Connectionless Data Transmission The connection and connectionless modes of data transmission are the fundamental models of communication within the OSI architecture. However, work on the OSI standards was well under way before the importance of the connectionless model of data transmission was recognized. Since both connection and connectionless data transmission modes can be applied to each OSI layer, addenda to the connection-oriented standards are being developed to cover connectionless operations.

See Figure 6-24 for possible layer-service combinations. Conversion from one combination to another is currently allowed, according to the Reference Model addendum which covers connectionless data transmission at the Network layer; it is possible at the Data Link layer if certain physical transmission systems are considered as connectionless. In other layers, the provision of one kind of service to the next-higher layer must always be accomplished by using the same kind of service from the next lower layer.

Connection-Mode Data Transmission. A connection proceeds through three distinct phases: establishment, data transfer, and release. No connection is established until all parties agree, through a process of

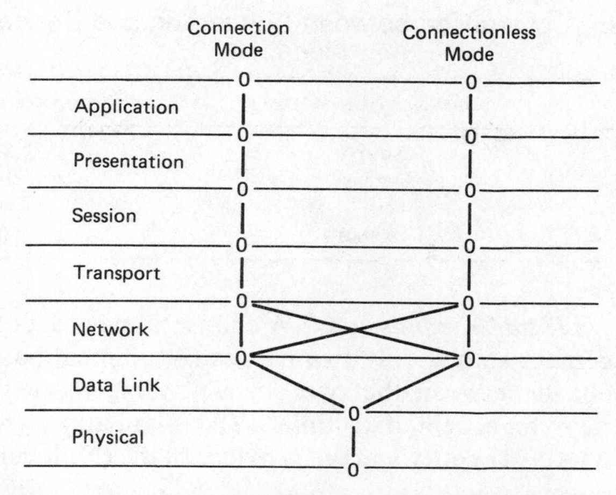

Figure 6.24. Transmission mode combinations. (*Source*: ISO DP 8524, Dec. 1984.)

negotiation, on the parameters and options that will govern the data transfer. An incoming request for establishment of a connection can be rejected if it asserts parameter values or options which are unacceptable to the receiver. The receiver may accompany its rejection with suggestion for alternative values and options. The negotiation process ensures that each party will reserve the necessary resources (e.g., buffers) to maintain the connection. During negotiation a variety of procedures for access-control, security, accounting, and authentication are carried out by the parties involved. In some instances agreements resulting from negotiation during connection establishment can be renegotiated during the data-transfer phase.

Three parties are involved in an (N)-connection: two (N + 1)-entities who wish to communicate and an (N)-service which provides the means to communicate. Once a connection is established it may be used to transfer successive data units until the connection is released by one of the three parties. Ordinarily, data units are related in such a manner that out-of-sequence, missing, or duplicated data units can easily be detected and recovered.

Flow-control techniques can be maintained to ensure that the peer-to-peer data transfer rate does not exceed that which the correspondents are capable of handling.

A connection identifier is assigned to a connection during its lifetime; data units carry this identifier rather than the actual addresses of the corresponding service access points. This technique reduces the overhead associated with the resolution and transmission of addresses.

Table 6.8. Comparison between Connection and Connectionless Mode.

	CONNECTION MODE	CONNECTIONLESS MODE
Agreement	3-party	2-party
Life cycle	3-phase	1-phase
Negotiation	yes	no
Data units	related	unrelated

Connectionless Data Transmission. A connectionless data transmission involves agreement among only two parties. Agreement has been established beforehand between the two corresponding (N + 1)-entities on their willingness to accept data units. The two-party agreements are between each (N + 1)-entity and the provider of the (N)-level connectionless service.

Connectionless transmission has one phase only. All the information required in order to deliver data units (destination address, quality of service, and various other options) is presented to the service provider along with the data in a single operation.

A transmitted data unit is entirely self-contained, and, from the service provider's viewpoint, bears no relationship to other data units. As a consequence, delivery of a series of data units in a particular sequence is not guaranteed. In some situations, however, there is a high probability of sequential delivery: for example, if the underlying (N − 1)-service provides a high probability of sequential delivery of (N − 1)-service data units, and all connectionless (N)-service requests are mapped onto a single (N − 1)-service access point.

Pros and Cons. Table 6-8 compares connection and connectionless transmission in terms of a few key attributes. Since in connectionless transmission a data unit includes all the information required by the service provider to deliver the unit to its destination, the connectionless service is more robust in a volatile environment. In addition, apart from the data units themselves, connectionless service permits the service provider to store less data. On the other hand, the larger data unit means greater overhead than incurred during the data transmission phase of a connection.*

Packet-Switched and Circuit-Switched Networks. Based on the connection and connectionless modes of transmission, communication networks

*Other benefits of connection-mode transmission include guaranteed delivery of data units in their correct sequence with provisions for recovery from lost, duplicated or corrupted data units.

fall into two broad categories: circuit-switched and packet-switched. In a packet- or message-switching network, fixed-length data-packets or variable-length messages are transmitted through a routing network from point to point. Such networks operate in a "store-and-forward" mode, in which the entire message or packet is stored at the various switching points until an appropriate path opens up. There is, therefore, never a clear and continuous path from input to output.

Circuit-switched networks, on the other hand, are based on direct circuit paths between input and output. These paths can be electrical connections or direct logical paths through gates. In circuit-switching networks data can be transferred directly from input to output at data-rates which depend solely on the electrical characteristics of the connection. However, the volume of traffic and the demand for system resources determine the ability at any given time to establish such a connection.

The telephone system uses circuit-switching, which is suitable for communications whose bandwidth requirements do not change much (as in the case of human speech), but which must be continuous. By contrast, terminal to computer or computer to computer traffic is usually bursty; much of the time there is no data to send, but every once in a while a burst must be transmitted. Packet-switching, which does not tie up expensive transmission facilities when they are not required, is best suited for this type of traffic, and most computer networks are of this kind.

In circuit-switched networks charges accumulate as long as the connection is intact; cost is proportional to the duration of the "call." In packet-switched networks lines are dynamically shared among all users on a demand basis; cost is proportional to the amount of data transmitted.

6.5.4 Carriers and Services

Throughout the world the electromagnetic spectrum is considered a public resource; the rights to supply communications facilities are therefore usually under government control. The suppliers of such services are either government agencies or private corporations operating under some form of government license. In North America the term "carrier" or "common carrier" denotes the supplier of communications services. In Europe the term "PTT" is commonly used to denote the governmental "post, telephone, and telegraph" authority.

Some carriers maintain their own transmission facilities; others lease bulk line capacity from other carriers, repackage this capacity, and market it to users. Not all carriers provide the same range of services, or charge the same rates for the same types of service.

In some cases facilities of different carriers can be interconnected under agreements among the carriers involved; in other cases, special arrangements must be made, either by the carrier at the point of origin or by the user, to complete a circuit which overlays the facilities of two or more carriers.

In the U.S. a relatively free competitive market for communications facilities encompasses six types of carrier:

1. *Telephone companies,* or *common carriers,* use voice facilities for data communications. This is the best established and most experienced group, in spite of the fact that their facilities are not always the most advanced technologically. Telephone companies must often provide the local loop, even when the longer-haul service is provided by a competing carrier.

In the U.S. the common carriers are AT&T, the Bell system of local operating companies, and more than 1500 independent local telephone companies. Interstate service provided by all telephone companies are regulated by the Federal Communications Commission; state authorities regulate services within individual states.

2. *Specialized common carriers (SCCs)* exist only in the United States, as a result of a 1971 FCC ruling to open up the supply of microwave-based leased-line facilities to independent companies. For the most part, SCCs offer voice, data, facsimile, video, and telemetry services in direct competition with the common carriers. Typically, an SCC constructs a microwave radio backbone linking two or more major population centers, but uses local facilities supplied by a common carrier to link its customers to the backbone. As for common carriers, SCC rates are regulated by the FCC for interstate services, and by the individual states for intrastate services. These rates are commonly lower than those charged by common carriers for comparable service. Most provide end-to-end service by purchasing local-access lines from the telephone company, and then billing their customers for the entire link. A good example of a specialized common carrier is MCI Telecommunications.

3. *Record common carriers* supply telegraph service. Western Union, the pioneer of American telegraphy, is the best known. It operates a nationwide network of facilities for voice, data, and video transmission.

4. *Satellite carriers* such as the International Telecommunications Satellite Organization (INTELSAT) offer intercontinental communications. There are also a number of U.S. domestic carriers in this field, such as Satellite Business Systems.

5. *Value-added carriers* purchase "raw lines" from other carriers, and add network management and error-control functions. Packet-switching on the basis of the CCITT X.25 standard is the most frequently encountered value-added service. Among North American value-added carriers are TYMNET, which is a virtual-circuit packet-switched network, and GTE Telenet.

6. *International record carriers (IRCs)* are licensed by national governments to offer services between specified countries. In most countries the international carriers are required to terminate their facilities in certain gateway cities, or at the facilities of the local telecommunications authority. IRCs provide telex, international telegram, and leased circuit services. RCA Global Communications and ITT World Communications are examples.

In Canada telecommunications are dominated by two major carriers, Telecom Canada (the Bell system) and CNCP Telecommunications. In most European, African, and Asian countries data communications are under control of a national postal and telecommunications authority.

AT&T and the Bell System The breaking up of the AT&T Bell System, which has been synonymous with telephone service in the U.S. for a century, has had a significant impact on the telecommunications industry. After 50 years as a monopoly regulated and protected by the government, AT&T agreed, in the face of a series of lawsuits, to divest itself of 22 regional telephone operating companies. The resulting AT&T system organization is depicted in Figure 6-25.

In data communications the new AT&T offers the following services:

1. *Switched Analog*. Unconditioned voice-grade lines which are subject to distortion caused by environmental interference and network traffic. This service involves several generations of telephone switching equipment. Three types of analog service are provided:

(a) *Long distance* provides public dial-up service for point-to-point communication among LATAs (local access and transport areas), over conventional switched lines. Its most attractive feature is its broad coverage, since the telephone system reaches about 150 million telephone stations in the U.S., is interlinked with the Canadian system, and can reach telephones in 115 other countries through direct distance dialing.

(b) *Wide area communications service (WATS)* is an outgoing call bulk rate service for high-volume users of the switched voice and data communications network. Several packages are available, based on the amount of time the service is used per day, and the geographical area covered.

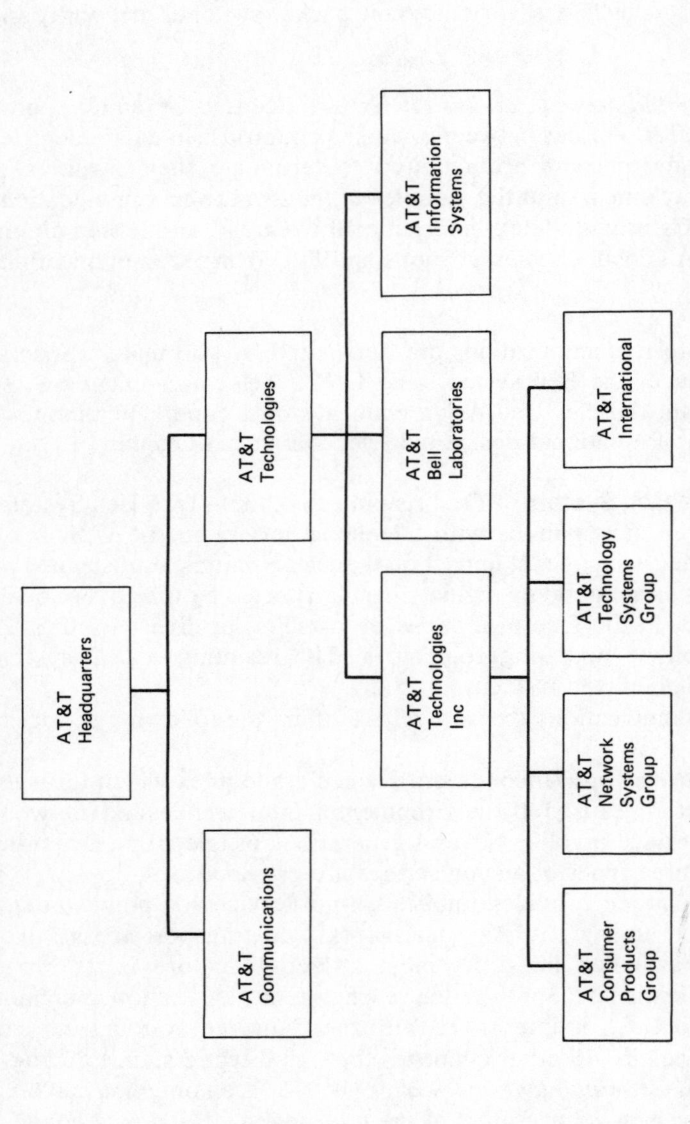

Figure 6.25. The AT&T system organization. (*Source*: The Auerbach Staff, "The AT&T and Bell Systems," *Data Communications Management*, REV 54-01-09.

(c) *800 service* is an inbound service, covering calls from outlying points to the 800 subscriber, that is, identical to WATS but in the opposite direction.

2. *Dedicated Analog Facilities*. These are private or leased lines, the most commonly used means of data communication. Although the transmission medium used to carry leased line traffic may be the same as that used for switched traffic, for dedicated services the routing and transmission quality are engineered to meet specified criteria. Dedicated lines are leased for full-time use. A choice of bandwidths is available; the greater the bandwidth, the higher the maximum speed of transmission. AT&T offers dedicated lines for services such as telegraph, voice, data, television, and facsimile. Its wideband service can support a channel speed as high as 230.4 kbps.

3. *Terrestrial Digital Facilities*. These are the Dataphone Digital Service and the ACCUNET services. DDS is a digital service for medium- and high-speed data transmission; it uses a "data-under-voice" technique on such media as the T1 carrier system and microwave facilities. The main advantages of DDS are its reliability and low cost.

ACCUNET provides high-capacity digital service over T1 lines operating at multiples of 56 kbps (as many as 24). It offers a packet service which is used in the development of packet-switched networks by individual customers, and an end-to-end circuit-switched capability.

4. *Satellite Services (SKYNET)*. The services provided are similar to the terrestrial offerings. In addition, however, SKYNET offers monochrome or color television transmission, and a stereo transmission service for audio.

The 22 Bell operating companies (BOCs) are controlled by 7 Bell Regional Holding Companies (BRHCs), which coordinate their activities through the Central Services Organization (CSO). See Figure 6-26 for the Bell system organization and Figure 6-27 for the BRHCs distribution. Initial offerings of the now independent Bell system companies complement the AT&T services. In addition, the operating companies have inherited the mobile phone system from the AT&T Cellular Company.

Telecom Canada Telecom Canada is a voluntary association of the ten largest Canadian telecommunications companies. Its purpose is to construct and maintain a national telecommunications network by coordinating data and voice facilities on a carrier-by-carrier basis. Telecom Canada offers some unique services.

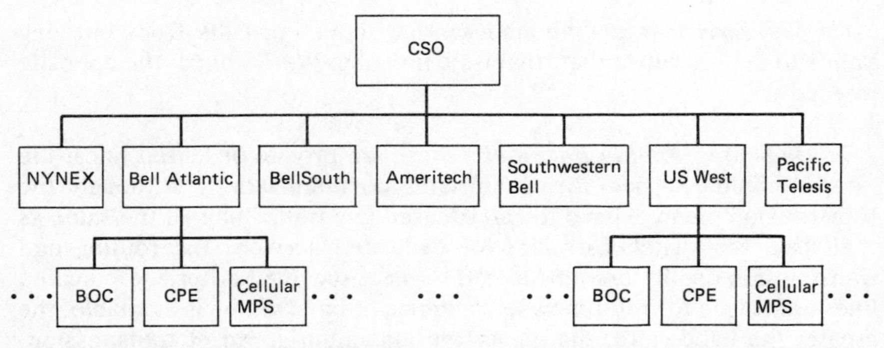

Figure 6.26. The Bell System organization. (*Source*: The Auerbach Staff, "The AT&T and Bell Systems," 1984, *Data Communications Management*, REV 54-01-09.)

Dataline II and III services transmit data from originating stations (users) to a centralized data terminal (a provider, such as a time-sharing computer), then return the processed data to the originating point. Calls cannot originate at the central data terminal.

Datalink is a pay-as-you-use, end-to-end digital circuit-switched data service. It is a transparent service which assumes compatibility between the communicating machines. Among its optional features are network security, local testing, and speed-calling.

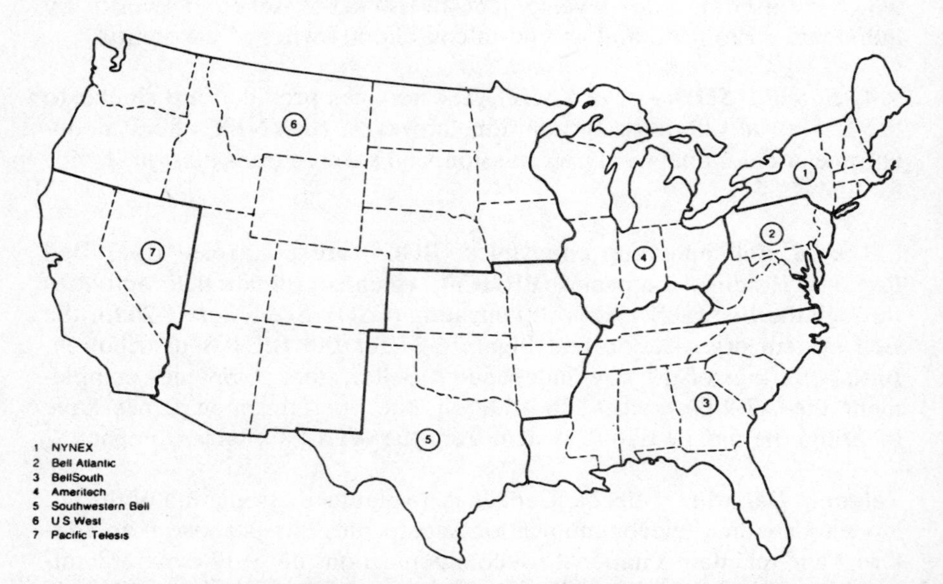

Figure 6.27. Regional holding company distribution. (*Source*: J. P. Heatherly, "Divestiture and Deregulation Issues," *Auerbach Data Communications Management*, 1984, 50-20-15)

Datapac is a packet-switched public data network which supports several different access services. Subscribers can interface to Datapac in either of two ways. The first provides access for packet-mode terminals or computers observing the X.25 standard.

The second gives access to non-X.25 terminals, such as teletypewriters and point-of-sale terminals. Such terminals, which cannot operate in packet-mode, connect to Datapac by means of packet assemblers/disassemblers (PADs). The PAD handles the protocol specific to the terminal, and performs a number of functions which support communications with a host computer connected to the network through an X.25 interface. Acting on commands from the terminal, the PAD establishes and clears all transmissions. During data-transfer the PAD assembles data received from the terminal into packets, and disassembles packets received from the host.

Telecom Canada offers a number of innovative value-added services.

Envoy 100, for example, whose terminals can be seen in Canadian airports is a national messaging service designed to accommodate the increasing flow of business information. It provides a wide range of customer-selectable features, allowing users to prepare, correct, send, distribute, access, and file messages to be passed between or within participating companies.

Users can access Envoy 100 through the telephone network, Datapac, the TWX network, or through the American GTE Telenet and TYMNET packet-switched networks. Message preparation and delivery options include line and text editing, delivery assurance, batch entry, staggered delivery in different time zones, commands in English or French, and user-defined storage capabilities for messages, distribution lists, or formats.

EnvoyPost is an electronic mail service offered by Telecom Canada in cooperation with Canada Post Corporation (the crown corporation which runs the post office). EnvoyPost uses Envoy 100 to access the national mail stream. Subscribers to Envoy 100 can send messages to non-subscribers in 5 million households and businesses in 20 cities across Canada. EnvoyPost prompts the sender for addressing information and text. The postal code information allows Envoy 100 to route the message to the appropriate Canada Post electronic mail center. The message is then printed on letter-quality bond paper, sealed in a distinctive envelope, and delivered on the same or the next business day, as the sender prefers. The U.S. Mail offers a comparable service, E-COM.

iNet 2000 is an intelligent network concept, developed to meet the growing need of managers and professionals for access to information sources and other computer-based services. The network is termed "intelligent" because it performs many computer access procedures which would otherwise require specialized knowledge of the user. Among its features are a directory of available services, automatic access to connected hosts, individual user-profiles recognized by the network, and consolidated billing.

6.6 NETWORK INTERCONNECTION DEVICES

Networks can be viewed as islands of automation. Network interconnection devices provide the means of connecting such islands (Figure 6-28). If all networks were built according to the same architecture, and followed the same protocols, then interconnection would involve only routing and naming management functions. Routing is required to transfer messages from one network to another; naming management is necessary in order to translate the names used in one network to address objects in another network using their correct representation in that network.

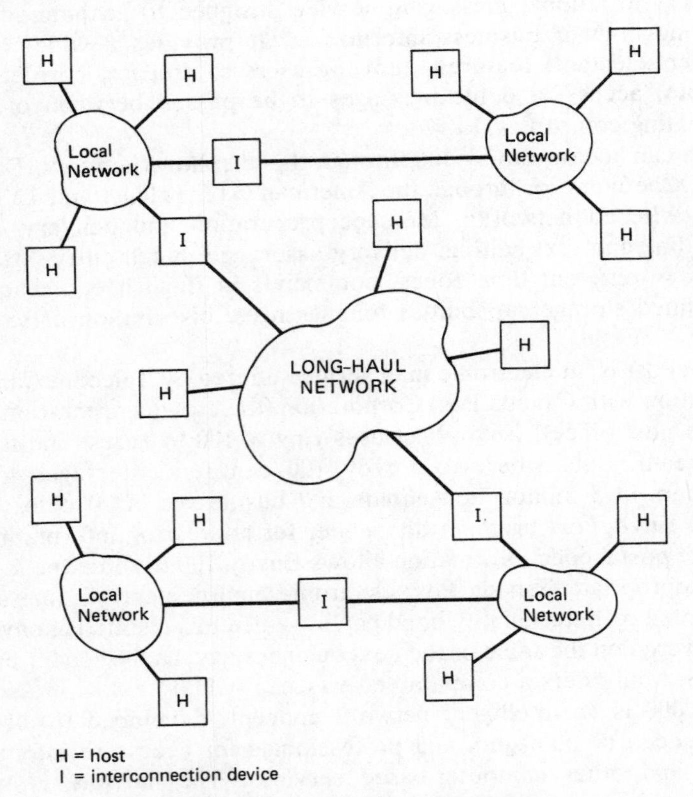

Figure 6.28. Interconnected networks.

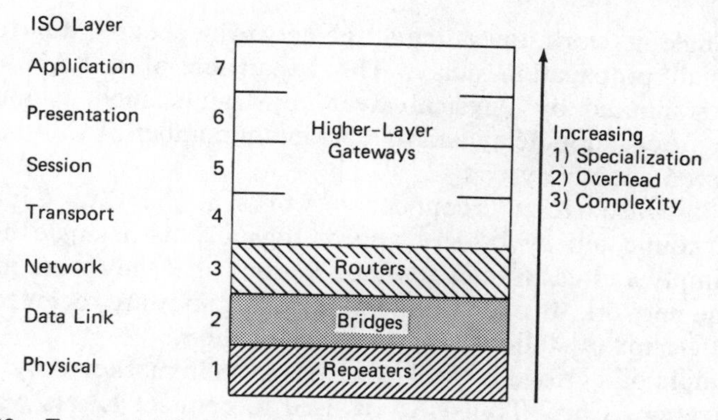

Figure 6.29. Types of interconnection devices. (*Source*: Hart, *Data Communications*, March 1985.)

However, since networks conform to different communications standards, interconnection devices must be introduced in order to convert protocols. Technically, the problem of interconnection is more complex when incompatible transmission media are used: for example, in interconnecting a 10 mbps coaxial cable LAN with a 56 kbps digital channel; the interconnecting device must exercise flow control, using buffering to avoid swamping the digital channel with traffic, and to avoid a bottleneck.

Related topics addressed elsewhere in this section are:

- The *private branch exchange* (PBX), which serves as a sophisticated switch for both data and voice traffic.
- The *metropolitan network,* which resembles an LAN in that it provides high-bandwidth over a limited geographical area. Unlike LANs however, such networks are owned by a common carrier.
- Facilities which allow users to *bypass* the local loop provided by the telephone company, and to connect directly to a long-haul carrier.

6.6.1 Repeaters, Bridges, Routers, and Gateways

Figure 6-29 associates various types of interconnection devices with the OSI Reference Model layer in which it relays messages from one network to another. The layer performing the relay does not use information provided by higher layers. Generally, the higher the relay layer the more specialized the products and protocols serviced by the relay. In addition, overhead and complexity increase as the level of the relay layer increases.

Repeaters relay Physical layer protocol data and control signals. They are used to extend LAN configurations by connecting channel segments directly or through an internal point-to-point link. Repeaters are used within a single network, operating at the network speed, and introducing only a small propagation delay. The expansion of a LAN through repeaters is limited by Physical layer constraints such as maximum round-trip propagation delay and the maximum number of stations which can be served effectively.

Bridges are used to interconnect networks at the Data Link level. Networks connected by bridges appear logically as a single network; stations simply address frames to other stations as if they were members of the same network. Broadcast and multicast frames are received by the addressed groups of stations regardless of location.

An example of a bridge is the TransLAN system marketed by Vitalink Communications Corp. TransLAN is used to connect LANs which use the IEEE 802.3 CSMA/CD protocol. Interconnecting LANs by means of either terrestrial lines or satellites, it operates by ''learning'' the identity of stations on the local network, and relaying frames to and from the local LAN. Frames which originate from and destined to the local LAN and are received by the bridge are discarded, as are frames received from remote LANs which are not destined for the local LAN. A TransLAN bridge supports as many as 8 satellite channels or terrestrial link ports, each with a capacity of 224 kbps. Frames are processed at a rate of 4000 frames per second, and forwarded at 1500 per second. Another bridge example is the interconnection of hyperchannel LANs through satellite links [Franta and Heath 1984].

Routers are the traditional way of interconnecting LANs. They operate at the network layer, and require that stations distinguish between local and remote communication. In order to communicate with a remote station, a station must send or receive packets to and from the router on the same local network. A special internetwork protocol is used at the network layer on each station. Routers use the internetwork control information in the packets, as well as a local configuration topology table, to determine how to relay a packet from the local to the remote network. Since stations perform filtering functions on behalf of the router, router-to-router links need not operate at local network speeds.

The ISO has a draft international standard which defines the internetwork protocol [ISO 8473 1986]; this is expected to replace the U.S. Dept. of Defence Internet Protocol, which has been the de facto standard.

Routers can interconnect networks which differ in their physical and Data Link architectures. For example Chay, Seltzer, and Siddique [Chay et. al. 1984] describe a router implemented with LSI chips which connects a CSMA/CD with a token bus baseband LAN.

A gateway operates in the higher layers (above the Network layer) of the OSI model, and interconnects users of different architectures. Since the protocols used to carry user data are different for each architecture, the gateway provides translation of the protocols from one architecture to the other. Gateways are visible devices in the sense that a station communicates with the gateway in order to use its services.

Gateways have a major role to play in opening up communications. Since there are so many communications standards (as well as many non-standard architectures), vendors tend to incorporate gateways in their own architectures to facilitate interconnection with networks from other manufacturers. The most commonly provided gateway leads to IBM network products, since IBM has the lion's share of the computer market. Any vendor who wishes to sell to IBM customers will need such a gateway.

One instance is the Internet set of products introduced by DEC to link their systems with IBM, UNISYS, and Control Data Corp. systems. Various IBM Internet products permit interactive block-mode (3271) and batch block-mode access (2780/3780). A DECnet/SNA Gateway allows cooperative computing on a two-directional link between a DECnet LAN and an SNA networking environment. Besides providing remote job entry, 3270 Terminal Emulation, and Applications Interface support, the gateway supports the Distributed Office Support System (DISOSS), a key component in IBM's office automation software. It also supports IBM's Host Command Facility, thereby allowing a 3270 user to access VAX systems through the DECnet/SNA gateway. Another relevant product offered by DEC is the DECnet Router/X.25 Gateway, which interconnects DEC's LAN to an X.25 packet-switched data network.

6.6.2 Metropolitan Networks

The last of the IEEE LAN standards to be considered is 802.6, which may someday be a specification for a network which can link devices throughout an entire city. A metropolitan area has a radius in the tens of kilometer range. Work on the standard began in 1982, but thus far no draft specification is available. Two proposals for the metropolitan network have come from 3M Interactive Systems and from AT&T's Bell Laboratories.

The 3M scheme is based on standard CATV video channels, as a medium for data transmission at 3 mbps. A centralized controller polls stations in turn, and retransmits each response. The time of an interrogation is a function of the physical placement of the polled station. Because of the relatively long propagation delay encountered in city-wide networking, the centralized controller times interrogations to minimize the idle time on the channel.

The Bell Labs scheme, called Homenet, employs what Netravali of AT&T describes as a novel CSMA/CD type of protocol, one which does not exhibit the usual loss of efficiency and fluctuations in packet-delay caused by collisions. The scheme is capable of handling both packetized data and voice on unused CATV channels.

Other media beside CATV coaxial cable may eventually find application in the metropolitan network; microwave is a possibility.

Another concept pertinent to metropolitan communications is the "teleport," an urban gateway providing efficient and economical communications services to long-distance users. The teleport is a communications hub surrounded by office complexes located just outside a major urban area. There are currently eleven operational teleports in the U.S., with two under construction and eight more planned.

The teleport idea started with a 1979 joint study, by the Port Authorities of New York and New Jersey and the National Research Council, on "Telecommunications for Metropolitan Areas, Near-Term Needs and Opportunities." The subsequent teleport project, based on a partnership between the public and private sectors, has four distinct and independent parts [Annunziata 1985]:

1. A satellite communications center on the teleport grounds (located on Staten Island). This will access all existing and future domestic and international satellites from a single location which has been cleared of radio-frequency interference.
2. A regional fiber-optics network spanning a large portion of the New York metropolitan area, and connecting corporate locations to the teleport site. This network is designed to handle up to 560 mbps at a bit-error rate of 1 in 10^9, with 99.99% system availability.
3. An office park development with a master plan of one million square feet of office space, to be constructed by private developers.
4. Enhancements to office park communication through constant monitoring and control of all teleport traffic.

6.6.3 Bypass

The concept of bypass emerged as a result of the divestiture by AT&T of the Bell Operating Companies, and the subsequent establishment of new definitions of local service. Bypass is an extension of the role of the specialized common carrier (see section 6.5.4). Just as the SCCs offer an alternative to the traditional long-distance telephone service, bypass provides an alternative for local calls: that is, within the local areas established for each of the new Bell Regional Holding Companies.

The divestiture of the operating companies has produced a new concept of local service, called LATA (local access and transport area). LATAs are large geographical areas that represent communities of interest; they also define the boundaries beyond which a long-distance carrier must be selected to transport calls from one LATA to another, even within the same state. The rationale for the LATAs is that they will ensure that BRHCs can continue to collect revenues from the stream of local calls. The LATA definition expands the local calling area in such a way that a substantial number of calls are placed at time- and usage-sensitive rates. Cost advantages once offered by intra-state WATS/800 services for calls within a region are no longer available.

Data networks which connect sites within a single region are affected by the establishment of LATAs. The number of sites considered to be within the local area (now the LATA) can be increased; but more and longer data loops may be required to connect a location which may previously have been served by bulk facilities.

LATA allows only one point of presence for each long-distance carrier; a local call must be made in order to reach the long-distance carrier. If the customer is located near the long-distance carrier's point of presence, little or no extra cost is incurred. If he is not, however time-, distance-, and usage-sensitive costs are added to the long-distance charge. Businesses with sufficient calling volume are given direct access to the long-distance carrier by means of facilities leased from the BRHC. These, however, have become more expensive under the LATA definition.

The bypass concept is intended as a competitive alternative to the Bell local loops; it would provide direct access to the long-distance carrier's point of presence. Bypass vendors use existing facilities which have been installed to support other services; they therefore permit the resale of excess capacity in facilities involving local loops. In theory, bypass can provide service at a lower cost than that of the Bell local loops, because the bypass vendor can both maximize the use of each facility and take advantage of economies of scale.

The main effect of bypass will be on metropolitan areas within LATAs; the principal customers will be businesses with high volume of communications. Bypass service requires easy access to the customer's premises, which is possible only in densely populated urban areas.

The following methods of bypassing the local loop are currently feasible:

1. Cable television (CATV)
2. Microwave and millimeter radio systems
3. Digital termination systems
4. Communications satellites
5. Cellular mobile radios
6. Fiber optics
7. Laser and infrared transmission

Bypass presents a major problem to Bell Regional Holding Companies, which face a considerable loss of revenue; such a loss would lead in turn to very high charges to local telephone service to residences and small business. Bell Communications Research has determined that the local telephone companies stand to lose about $6 per access to bypass [Hayes 1985].

6.6.4 Private Branch Exchanges (PBXs)

A PBX is a device which connects users of single-line telephone instruments to other users either inside or outside their own organization (Figure 6-30). A typical PBX uses twisted-pair wires to connect all devices. It has a star configuration. When a device wants to establish a connection, it signals the PBX. Upon acknowledgment, the device sends a destination address, and the PBX uses its switching matrix to complete the connection and establish a session.

The original private exchanges were switchboards requiring an operator to make and break all connections. During the 1920s these were replaced by automatic systems called private automatic branch exchanges

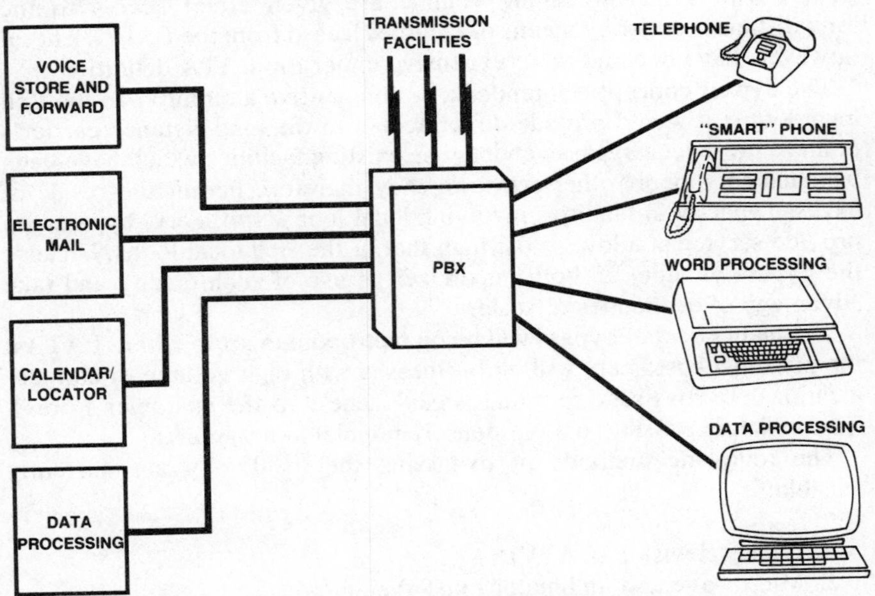

Figure 6.30. PBX user interconnection. The PBX functions primarily as a circuit switch connecting the user with specialized processors for voice store-and-forward, voice recognition, and electronic mail, as well as with data processors. Speed, code, and protocol conversions may also be provided, depending on the processor's capacity. (*Source*: G. M. Pfister and B. V. O'Brien, "Comparing the CBX to the Local Network—and the Winner is?" *Data Communications*, July 1982, pp. 103–113)

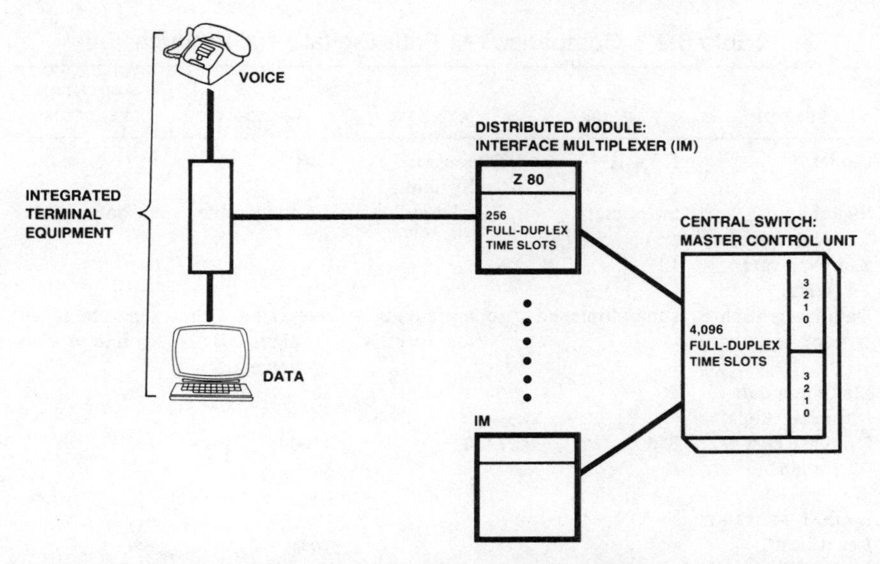

Figure 6.31. Third generation PBX. The three basic characteristics of the third-generation PBX are a distributed architecture, integrated voice and data by design, and a non-blocking configuration. Shown here is one PBX example, Intecom's IBX. The integrated-terminal-equipment device includes a digitized telephone. (*Source*: G. M. Pfister and B. V. O'Brien, "Comparing the CBX to the Local Network—and the Winner is?" *Data Communications*, July 1982, pp. 103–113)

(PABXs), which used electromechanical "crossbar" switching technology and analog signalling.

In the mid-1970s the digital PBX, or computerized branch exchange, was introduced. These PBXs use electronic technology, with digital internal switching. They were initially designed mainly to handle analog voice traffic, but embody a codec function to permit internal digital switching. These systems were also capable of handling digital data connections, without the need for a modem.

Current or third generation PBXs are touted as "integrated voice/data systems." They provide the following enhancements (Figure 6-31):

- the use of digital phones in integrated voice/data workstations
- distributed architecture involving distributed control of multiple switches
- non-blocking configuration, with dedicated port assignments for all attached phones and devices

Analog and digital PBXs seem to be equivalent in the sense that the analog PBX handles telephone sets directly and uses modems to accommodate digital data devices, while the digital PBX handles digital data

Table 6.9. Comparison of Four Digital PBX Products.

VENDOR	ROLM	AT&T IS	INTECOM	NORTHERN TELECOM
Model	CBX-II	Dimension System 85	IBX	SL-1
Switching technique	proprietary	T1 compatible	T1 compatible	T1 compatible
Sampling rate (KHz)	12	8	8	64
Data integration technique	submultiplexed	separate data line to PBX	integrated with digitized voice	separate data line to PBX
Maximum data speed (kbps)	56	56	56	56
Purchase cost per station* ($)	850	1100	1100	900
SPECIAL FEATURES				
Least cost routing	yes	yes	yes	yes
Automatic call distrib.	yes	uniform call distrib.	yes	yes
Call restriction	yes	yes	yes	yes
Traffic measurement	yes	yes	yes	yes
Call accounting	yes	yes	no	yes

*500 station system; all stations with voice capability, 25% with data capability.

SOURCES: 1. Levin, Auerbach Data Communications Management 1984, 53-10-21.
2. Datapro Research Corp., Aug. 1984, 612-010-101.

devices directly and uses codecs to accommodate telephone sets. The digital approach, however, offers some advantages:

- Digital technology lends itself to software and firmware control, and to the use of VLSI components.
- It facilitates time-division multiplexing, which makes efficient use of internal data-paths and switching techniques, as well as providing access to public carriers using time division multiplexing.
- Digital control signals can be easily integrated into a transmission path by means of TDM.
- It accommodates encryption more readily than analog techniques

Table 6-9 compares four leading digital PBX products.

The PBX is the most cost-effective voice network, and its ability to handle digital data makes it an alternative to standard LANs. Data and voice can be combined either at the telephone set or in the PBX. In the former case,

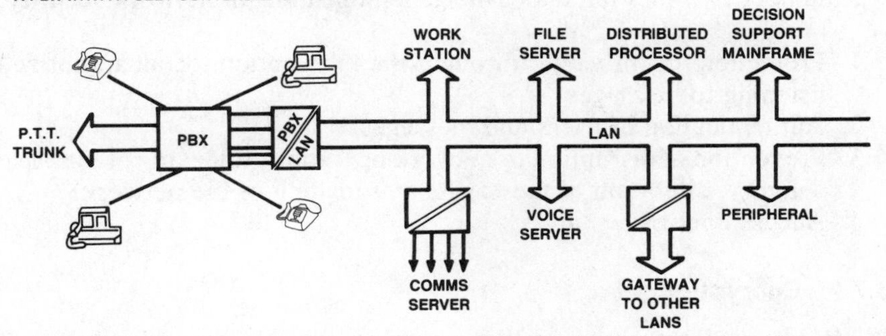

Figure 6.32. A PBX with a fully converged office system. (*Source*: J. Houldsworth, "Convergence of LAN and Digital Telephone Exchange Systems," NATO ASI Series Vol. F6, *Information Technology and Computer Network* (Edited by K. G. Beauchamp), Springer-Verlag, Berlin Heidelberg 1984, p. 151)

digitized voice is time-division multiplexed with data onto the telephone cable, then separated into two paths at the switch. In the second, analog voice and digital data travel through separate cables from the telephone location to the PBX. The Rolm CBX-II and Northern Telecom SL-100 use the first method; the Northern Telecom SL-1 uses the second.

The LAN and PBX are not necessarily competing products. A recent development is the integration of the LAN and the PBX in, typically, an office system (Figure 6-32). Such a configuration combines many benefits of both, such as high data rates among servers, and voice and data switching capabilities.

Another new development is the digital multiplexed interface (DMI). DMI is a high-bandwidth multichannel computer-to-PBX interface which uses a T1 (in North America) or CEPT (in Europe) carrier. DMI is an AT&T product and a proposed standard which is viewed as a first step toward economical interfacing of PBX with ISDN.

6.7 COMMUNICATIONS AND NETWORK SECURITY

The transmission mechanisms used in communications are vulnerable to intrusion of two kinds: a passive intruder merely listens to communications, while an active intruder alters or inserts messages, or retransmits valid messages. Both types of intrusion can be accomplished by wiretapping (that is, by making an illicit physical connection to a communications path). Apart from the transmission channels, vulnerable points exist at switching centers and at interface points between host machines and the network. In general, long-haul networks are more vulnerable to security breaches than LANs; they cover wider areas, and often use public carriers.

A number of issues fall under the general heading of network security:

1. Protection of privacy, through the prevention of unauthorized listening to messages
2. Authentication of users and messages
3. Prevention of disruption of network operations (blocking of message delivery, alteration of messages, overloading of the network)
4. Access control

6.7.1 Encryption

While physical measures, such as burying cables, can help, the most important security technique is encryption. Encryption of data can be carried out at several levels. The most efficient point is just before the message goes out on the communications line; at the Data Link level, where encryption can be carried out along with other manipulations such as check-sum calculation. The problem here is that either a single key must be used for all communications between the two network nodes, or some authority must be entrusted with the keys for all users. In addition, message security depends on the correct functioning of all the levels of system software between the user and the communications line. If any intervening switching node is subverted, all traffic passing through that node is exposed.

In end-to-end encryption the key is chosen by the user (or the application) and is otherwise known within the system only by the encryption mechanism itself. If there is a chance that the key is compromised, then it can be changed. In end-to-end encryption the choice of security measures can be left up to the users, who will bear the added cost [Voydock and Kent 1983].

Private Key A key is required for each potential pair of communicators. A large number of keys must therefore be distributed in some secure way. One approach is to establish a key distribution center, which maintains all keys and, through a special key, communicates these as required to all nodes. When a node wishes to communicate with another, it asks the key distribution center to send keys to both parties. One obvious problem with this arrangement is that the system fails if the distribution center fails or becomes congested. A refinement is to distribute the key distribution function among all nodes.

Public Key In a public key encryption system it is necessary to maintain a correct and up-to-date listing of public keys. All changes to the listing must be authenticated; a sender who is given a public key for a potential receiver must be sure that it is correct [Popek and Kline 1979]. One problem with this type of system lies in the authentication of users: passwords communicated in plain text can be compromised by passive intrusion. A solution is to have the system call the user back at a privately listed telephone number.

Digital signatures are another possible way of authenticating users and messages. A digital signature must have a number of the attributes expected of a written one. It should be very difficult to forge; the receiver should be able to validate the signature when the message is received, and should be able to demonstrate at a later time that a valid signature was received. Both public-key and central authority schemes have been proposed as bases for digital signature systems.

6.8 NETWORK PERFORMANCE

As the open system concept achieves widespread acceptance the number of terminals connected to networks, the number of users seeking access to host machines, and the volume of machine to machine traffic will increase. In such an environment network performance will be of prime importance. For example, patterns of access to hosts will be "bursty," because of factors such as communication over multiple time-zones; a problem which can be solved by adjusting the cost of communications according to the time of day, in the manner of the telephone companies.

6.8.1 Performance Factors

One of the main objectives of data communications is to provide fast data transfer. Data transfer time can be broken down into three components:

1. Computation time as the message is prepared for transmission
2. Transmission time, which measures the number of bytes per second which can be put on the medium, a function of the medium's bandwidth
3. Propagation time, which measures the time it takes the message signals to travel through the medium, which is a function of the medium and the distance the signal must travel

In order to reduce computation time many communications functions are performed by special hardware, which can carry out specific functions at high speeds. For example, the chips produced to perform codec and modem functions. The trend is to embody as many functions as possible in hardware.

Computation time necessary for data transmission is considered overhead for the communicating host machine. Therefore, front-end machines dedicated to the task of communication on behalf of the host machine are often used. Tailored to communications tasks, the front-end machine is normally connected to the host by a high-bandwidth channel. A further advantage of separating communications from other functions normally carried out by the host is that it permits changes in certain communications functions without affecting the host. In addition, the front-end can be brought down for maintenance independently of the host machine.

Propagation time is affected by the number of hops between source and destination: the number of nodes the message must pass through on its way. If a transmission from point A to point C passes through point B, then computation, transmission, and propagation delays will occur between A and B, and between B and C.

Propagation time is an important factor in satellite communication. For an earth distance of 1000 miles between communication points, the approximate propagation time over telephone lines is 0.01 seconds; for satellite communication between the same two points the propagation time is about 0.3 seconds. The problem of propagation time is also more important in long-haul networks which have lower bandwidths compared to LANs.

Since the volume of data transferred influences transmission time, there is always the danger that the transmission medium may be "taken over" by a single high-volume transmission (such as a file transfer), and that other services will therefore experience unacceptable delays.

Another common problem in communication is the ability of the receiving end to absorb the messages sent to it. The receiving end may be swamped by the sudden arrival of a massive volume of data; if it cannot process this data quickly enough, the data may be lost. Such situations are dealt with by a combination of throttling and buffering.

Throttling mechanisms enable the receiver to signal the sender to cease or to continue sending messages. An upper limit is placed on the number of messages the sender may transmit without receiving an acknowledgment from the receiver; when that bound is reached the sender anticipates a "stop" or "continue" message. Buffers at the sending, receiving, and intermediate nodes are provided to accommodate messages in transit. The number and size of buffers at the receiving and intermediate nodes determine the maximum number of messages a sender can transmit without acknowledgment.

6.8.2 Performance Monitoring

Network performance monitoring is the gathering of information about the performance both of components and of the network as a whole. Its purposes are:

- to measure the network against its performance specifications
- to control the load through the network in order to increase performance
- to locate and diagnose problem areas
- to tune the network to achieve specified performance
- to provide input for modelling the network for the purpose of performance analysis and prediction

Monitoring is carried out by both hardware and software tools. A hardware monitor typically consists of a set of high-impedance probes, a logic plugboard, a set of counters, a storage medium, and a display. The probes can be attached to memory locations, registers, and bus connectors. Through the logic plugboard the number of transitions of a given type is measured; for example, the number of clock cycles during which a processor is busy, or the amount of time the processor spends at certain priority levels. Such measurements can be directly translated into performance metrics such as CPU utilization, storage area utilization, or device seek/count time.

Other hardware monitors are used to collect and display statistics related to the movement of data on the communications line. These monitors evaluate such characteristics as signal form, missing bits, modem functions, and address formats.

Software monitors are processes which execute on the machine which is being monitored. They therefore consume system resources, and may have an effect on the performance measurements taken. The advantage of a software monitor is that it has access to the dynamic internal data structures of the system; it can provide statistical data on internal queues and on dynamic buffer allocation/deallocation.

6.8.3 Performance Modelling

The prediction of performance is a process by which, on the basis of the performance of a subset of observed samples, the performance of the other, unobserved subset is projected. Given a proposed network and its requirements, it is often impractical to set up the proposed network in order to determine whether it meets the requirements. Performance prediction tools offer a simpler means of determining, within a certain range of accuracy and confidence, whether the proposed network will do the job.

Performance prediction is required:

- when a new network is proposed
- when a network requires tuning

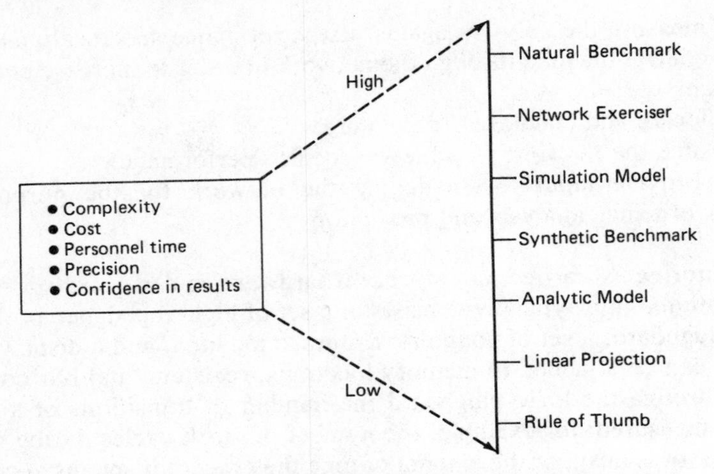

Figure 6.33. Spectrum of performance analysis techniques.

- in capacity planning, when the hardware, software, workload, or applications of an existing network are to be modified
- in identifying problems in the operation of a network

Figure 6-33 offers a general outline of techniques used in predictive performance analysis; the techniques are classified by degree of complexity, cost, and precision.

We now discuss briefly the advantages and disadvantages of each of these techniques.

The advantages of rule-of-thumb are that it requires relatively little effort—it is based on experience—and that it is unpretentious. The future workload of a network is often very difficult to predict; a more scientific technique might merely lend credibility to faulty projections. Its disadvantage is that rules-of-thumb are only educated guesses which could be wrong—they don't carry much weight as a scientific technique.

Linear projections are relatively easy to construct, and they can be fairly valid for a static environment depending on the level of detail and the amount of historical data available. On the other hand, networks are usually dynamic in nature and real-life relations among parameters are usually nonlinear.

The results of queuing models can be obtained quickly; they are efficient, inexpensive, and fairly accurate, and they provide real insight into network behavior. Moreover, network changes can be tested on the model prior to implementation. The disadvantages of this technique are that it depends heavily on assumptions about system behavior which may be wrong, that the calibration and validation of the model is often difficult, and that operation research specialists who understand both the network and the mathematics of queuing theory are required to build the model.

Simulation can provide fine and accurate results—for example, histograms of response times. Moreover it permits the analysis of performance in detail before a network is implemented or modified. It is, however, both costly and time-consuming, and for good results accurate and sometimes detailed information about the network is required. Such information is often difficult to obtain, as well as information required to validate the simulation model.

Benchmarking determines whether a network can adequately process a specified workload. It selects those configuration options and software parameters which will perform best, and demonstrates that the proposed hardware and software will actually work. It is, however, no easy matter to represent either a present or a future workload. Benchmark models are very expensive to develop, and moreover require a controlled environment with hardware and software already in place.

6.9 NETWORK MANAGEMENT

A network management system assists in the creation and testing of the network, and the performance monitoring and reporting of its activities. Network management activities are carried out by the network administrator who has special privileges for invoking network service operations. A special network operator interface handles commands received from the network administrator, and invokes appropriate services.

6.9.1 Performance Monitoring and Reporting

Monitoring is the gathering of statistics on network activity, normally at each communications layer and at each node in the network. This task is carried out automatically once the network administrator has determined the degree of monitoring required—since monitoring consumes resources and introduces overhead, full-scale monitoring may not be desirable. The logging of statistics can be carried out at the node in which the statistics are gathered, or the records can be transferred to be logged at another node. Statistics can be used for debugging, tuning, billing, and analysis of the performance of various network components. When the statistics indicate an alarm condition, this condition is displayed at the network administrator's console.

Monitored data can be forwarded directly to a network control center for on-line display of the current network status (dynamic statistics). Such a display can embody graphics, providing a pictorial representation of network components and relevant current information, such as load on the various nodes, and traffic volume on the lines. Alternatively, the log file data can be used to produce statistics for analysis and report generation.

6.9.2 Network Configurator

The network configurator includes a database of configuration parameters for individual nodes in the network and for the network as a whole. It has an interface which interactively assists the system administrator in configuring the network and each individual node, ensuring consistency of network parameters through validation checks. A node which does not include a network configurator may have a basic configuration which will allow it to connect to and download a configuration file from a node which does include a network configurator. The configurator database can also serve as a source of static statistics about network configuration.

6.9.3 Loopback Testing

Loopback testing enables the testing of physical and logical links, for example, through nodes, lines, modems, and communication interfaces.

Chapter 7

Distributed Databases

Computer applications are frequently built around data representing some "real world" state of affairs, stored on magnetic disk or a similar medium, and retrieved and/or updated by various programs as that state of affairs is analyzed or altered. The development of general-purpose tools to organize and manage data is an on-going topic of theoretical interest in computer science, and one of the major activities in the computer software industry. The current solutions to this class of problem are not subordinated to a coherent, consistent theoretical model even for closed, centralized systems; they are a patchwork of theoretical models and practical techniques applied as they fit specific cases. It is therefore not reasonable to expect that a single conceptual model will soon emerge to serve as a basis for the solution of these problems in open systems. The development of open systems must rely initially on models which classify and connect existing data management techniques. In order to understand this process we need first to consider the nature of data management.

The simplest organization of stored data is the flat file: a set of data elements accessible in sequence, one after the other, or by relative position. The data elements of a file may be simply atomic storage units, as in UNIX files which are a named array of bytes, or they may be structures, as in some MVS or VMS files which are named arrays of records, each having a length and an inner structure. In either case, the details of the structure and the semantics of the elements are not part of the file itself. The records of a given file refer to instances of one type of "real world" entity, with the attributes that describe and identify that instance encoded as values in fields of the record. The definition of the entity type, its attributes, and their domains and meaning, is stored and managed separately from the file, perhaps only as a record description within the programs that use the file, or perhaps as a data in another file that serves as a data dictionary. This descriptive information, data about data, is called *metadata*.

Individual files are often organized in more complex ways for efficiency reasons: the records may be ordered on the values of some attribute, and possibly arranged into a tree structure for faster searching; or the records may be distributed through the file space by a hashing function on the value of some attribute for very fast access; or the records may be stored in order of arrival and accessed through companion index files built from the values of various attributes for multi-attribute searching. In any case, the programs that manipulate the file directly must "understand" the organization and use the appropriate access methods. Further, the entities represented in one file are often related to the entities in another as part of a more comprehensive conceptual structure. Such records are often linked through common attribute values, as for example the item numbers in an inventory file might appear in the detail records of a customer's order file to identify what has been ordered. Here too, the programs that manipulate the files directly must explicitly know of the linkages that express the relationships and use the appropriate access methods.

It has long been recognized that building applications directly upon files leads to a number of problems. First, programs are harder to write and debug because they include explicitly the details of the physical data structures, file organizations and relationships. Second, the programs are harder to maintain because they must be changed to reflect any change in a file's organization or content. Where many programs are involved, dependencies may be overlooked, leading to inconsistent results in operation. A layer of software can be constructed between the files and the application programs that use them to provide a uniform conceptual view of the data, independent of the underlying physical structures and file organization, and to mediate between the conceptual access requests of the program and the file access methods. Such a layer of software and the files it manages is called a *database*.

This chapter will deal with the role of database management in open systems. Files are not unimportant in open systems, but their use is largely restricted to program development, where files are usually small and self-contained, and to applications involving bulk data transfer. Distributed file systems have been discussed briefly in Chapter 2, directory services for distributed file systems in Chapter 4, and protocols for remote file access and transfer in Chapter 6. The use of data across systems requiring coordinated action to ensure data integrity and consistency among multiple users is essentially a database problem.

A database has two principal components: first, the data stored in it, organized logically in terms of some conceptual data model and physically in terms of some set of storage and access techniques; and second, the metadata that defines the logical organization of the data and maps it to the physical structure. It is the stored metadata, often called the database catalog, the data dictionary, or the database schema, along with the software that uses the metadata to provide the high-level data manipulation, that distinguishes a database from mere files.

A *database management system* (DBMS) is the software that manages both the metadata and the data under the rules of some data model and some storage and access techniques. The term usually implies a generalized software package that allows an application designer/programmer to define and create independent databases, with a *data definition language* (DDL) for operating on metadata and a *data manipulation language* (DML) for operating on data. A *database system* (DBS) is an instance of one or more databases under a given DBMS.

A *centralized database* (DB) is one in which both its data and metadata are stored on and managed by a single computer system (a single-processor machine, or a multi-processor machine with shared memory or otherwise tightly coupled). A centralized DBMS is one that can create and manage only centralized databases.

A *distributed database* (DDB) is one in which its data or metadata (or both) are stored in parts or in multiple copies (or both) on more than one computer sytem, where the systems are connected by a communications network, and the database is managed in a coordinated fashion as a whole entity. A *distributed database management system* (DDBMS) is one that can create and manage a distributed database.

The concept of a distributed database grew not only out of the concept of open systems, but also out of the need to impose the discipline and control of a database across multiple computer systems within a single organization. In any given instance of its use, the term "distributed database" may describe one of a variety of situations; from a simple case in which the content of a single database (single in the sense that it has one global data definition and one overall administrative control) is distributed over several different processing sites in a network, to very complex cases in which several individual databases (individual in the sense that each has its own data definition, possibly in different data models, and its own administrative control, as well as separate locations) are to appear as a single database to some application or user. (Perhaps we should use the verb "to cleave," which means both "to split apart" and "to adhere together," so that distributed databases formed by fragmenting a unified structure might be called *cloven* databases, and those formed by unifying a fragmented structure might be called *cleaved* databases.) A common name is appropriate, however, because most of the problems to be solved in the implementation of a distributed database are common to all configurations, the extremes as well as intermediate cases. Nonetheless, the solutions take on different forms in different configurations, and some of the problems (e.g., translating between different data models) arise only in the more complex cases.

There has been considerable research on distributed databases, but the work is sometimes difficult to interpret and evaluate because of this breadth of context. The difficulties are exacerbated by the lack of a generally accepted taxonomy of the database management systems. Data models, which provide an implementation-independent framework for describing the conceptual operation of a database management system, are well developed but say nothing about the structure of the components of a DBMS or about the flow of data and control within it. Many of the problems encountered in distributed databases are interior to the DBMS, and must be understood in terms of the overall architecture, rather than just the data model. Several models have been proposed for the centralized DBMS, and some attempts have been made to extend them to the distributed case. However, there is no model of the DBMS with the completeness, clarity, and acceptance of the ISO/OSI model of intersystem communication. In Section 7.1, we review the work in this area and establish a classification scheme for distributed database management systems which identifies, at least for the purpose of exposition, several progressively more general orders of distribution.

Though a number of distributed database management systems have been designed, only a few have been implemented to date; and fewer still are being used in production (as opposed to research) environments. It is therefore difficult to assess the practical merit of some of the theoretical design proposals. In addition, much of the research to date has been directed at questions that appear technically challenging, but which may not be relevant to the real problems. The practitioner often has to solve hard, if uninteresting, problems with little guidance. Jim Gray, a practitioner who has also contributed significantly to the theory in this area, has stated:

> "Database theory seems to lag behind or be orthogonal to the practical problems of database design and implementation. In my own specialty of transaction management algorithms appear in working systems five years before they appear in the literature. The best theoretical papers explain the algorithms, expose subtle bugs, or generalize them." [Gray 1985]

As well as serving to caution the reader, this fact has guided our selection and organization of issues to consider in this chapter. In Sections 7.2 through 7.7 we examine the major practical problem areas found in centralized database management systems, and explore how these problems appear in the various classes of distributed system established in 7.1. These problem areas are:

- Data models (7.2)
- Conceptual design (7.3)

- Query processing (7.4)
- Transaction management (7.5)
- Physical design and implementation (7.6)
- Database administration (7.7)

Two problem areas are present only in distributed databases. The principal problem arises only in the more complex cases, namely that of providing a general-purpose directory or catalog of databases. This is discussed in Section 7.8. The second problem area is the interface between the DBMS and communications, which is discussed in Section 7.9.

7.1 DBMS ARCHITECTURE

In order to identify those aspects of a database management system that can be distributed, it is necessary to establish some general framework in which to describe the architecture of a DBMS. Though there is considerable diversity among DBMSs as implemented, there is also considerable commonality in that they are for the most part all intended to perform similar functions in similar operating environments, in spite of differences in data abstraction concepts. A number of attempts have been made to formalize this commonality: Slonim [Slonim et al. 1979] analyzed the functions provided by a number of DBMSs to establish a hierarchical representation of DBMS architecture; Brodie [Brodie 1980] used data abstraction concepts to represent DBMS architecture as a matrix of abstract objects and functions; and Date [Date 1980] has proposed a unified database language which combines the features and functions of the three standard data models.

An architectural framework should identify all data objects—including both the data managed by the DBMS and the metadata used by the DBMS to define a database instance—and should group them with manipulation functions into DBMS components. It should also identify the internal and external interfaces. A layered approach, as used in the ISO/OSI model of intersystem communication, is especially suitable to this task if the architectural model is to be independent of any particular data abstraction model. As part of an effort to lay a foundation for DBMS standards the ANSI-X3-SPARC Committee on Database Systems Management, proposed a layered model that identified external interfaces in terms of user roles. The upper layers are shown in Figure 7-1.

Three levels of schema (i.e., database description) are defined: an external schema provides a representation of a subset of the database for end-users and application programs; the conceptual schema provides a representation of the entire database for the database and application

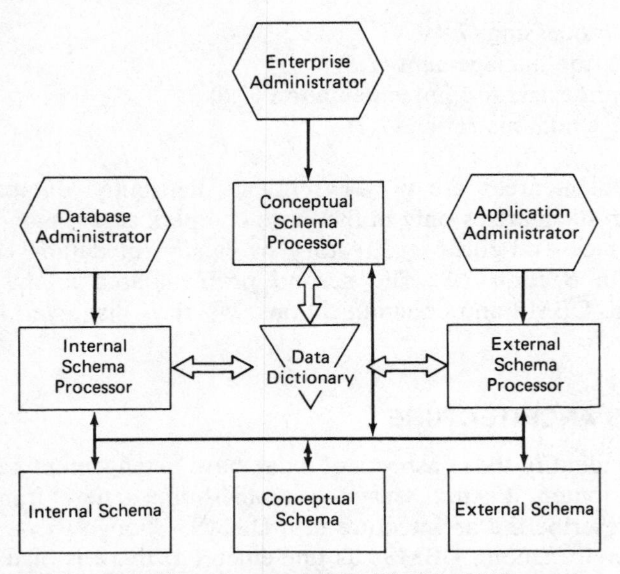

Figure 7.1. Upper layers of ANSI SPARC DBMS model.

administrators; and the internal schema provides a representation of the storage structure and access mechanisms for the system programmer and database administrator. Each schema level is used at compile-time and/or at run-time to map from the external view of the database to the actual stored data. (The lower layers of the ANSI/SPARC model define the run-time components that perform the actual data manipulation and storage management).

In a distributed environment it is necessary to extend this model to include a representation of how the database is distributed, and a mapping from a unified external view through this representation to the data itself. Several alternative extensions have been proposed. Devor et al. [1982] adds two layers of schema *between* the conceptual and the internal layers, a global representation schema and a local representation schema, and one instance of a local internal schema for every site holding part of the database. Bachman [Bachman and Ross 1982] merges the ANSI model with the ISO/OSI model as shown in Figure 7-2. The schema layers appear to have vanished, but in fact have been recast as data used by the functions in other layers. Larson [Larson and Rahimi 1984] adds a *distribution schema* that may be positioned anywhere between the other schema layers, and adds an optional sublayer to the external for a federation schema, as shown in Figure 7-3.

None of these extensions covers all conceivable cases, but Larson's seems to provide the most general framework in which to describe the

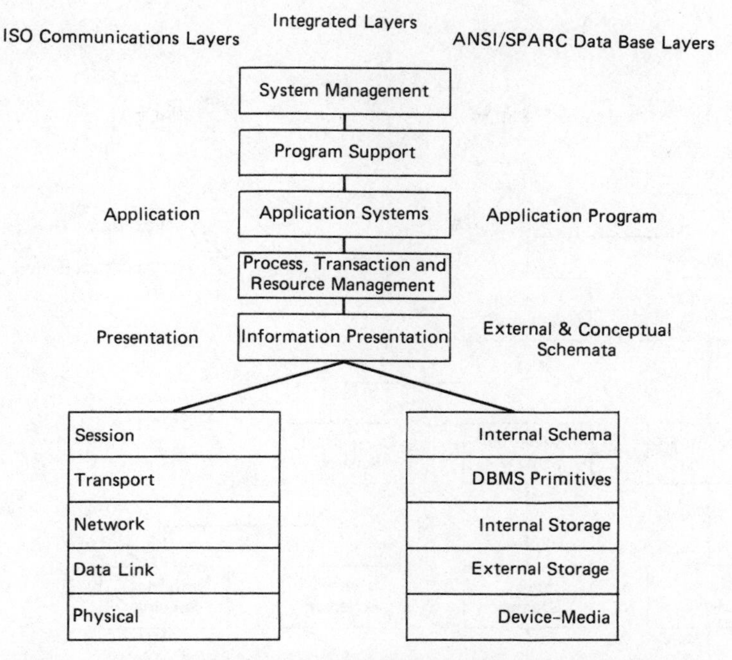

Figure 7.2. The Bachman integrated communications/DBMS model.

main classes of distributed database so we will adopt it as a framework for discussion.

Consider the aspects of a DBMS, in terms of the ANSI/SPARC model, that may be distributed (i.e., separated in structure but combined in operation):

- The data itself may be separated over multiple processing sites, but "combined" through one of the schema levels.
- The roles of those users who exert control over some aspect of the database (e.g., database administrator or application administrator) may be separated across organizational boundaries, but "combined" through one of the schema levels.

There are two major modes of data distribution:

1. *Partitioning*. Data is separated according to rules that allocate specific data to specific sites. *Vertical partitioning* assigns data to sites by conceptual definition; e.g., some defined attributes of relation R are assigned to site A, others to site B. *Horizontal partitioning* assigns data to sites by the value of some attribute in each instance of a record or tuple; e.g., tuples of relation R with a given

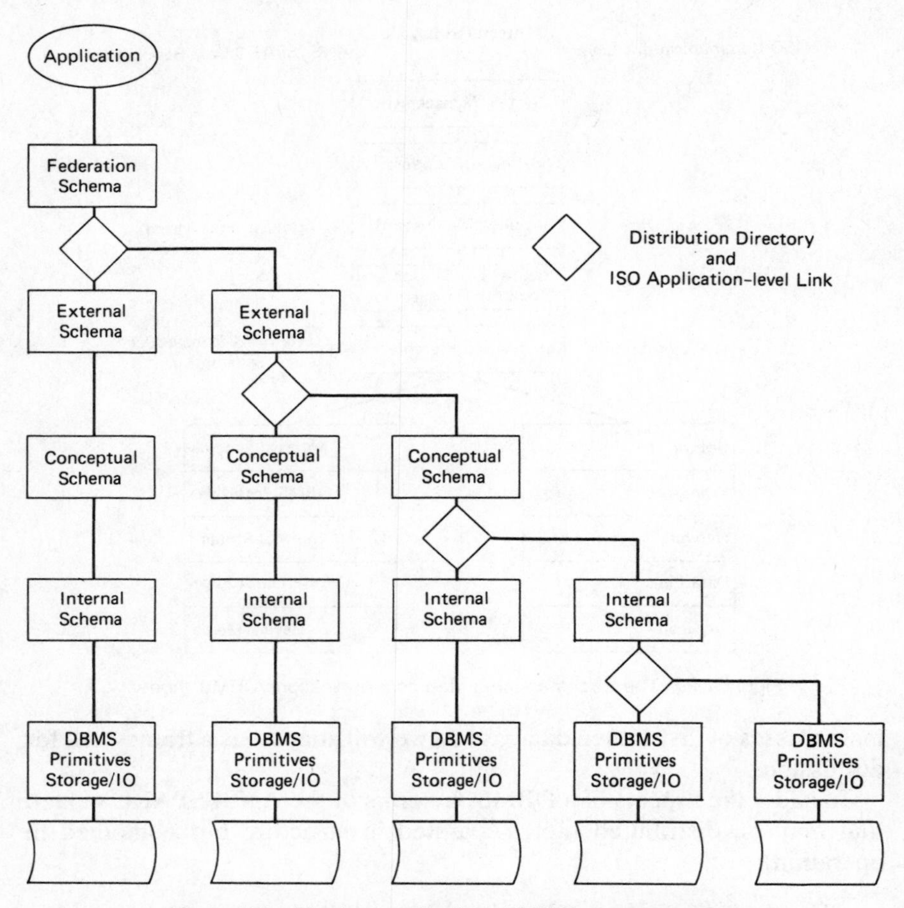

Figure 7.3. The Larson distributed DBMS structural model.

value of attribute X are assigned to site A, those with different value to site B, and so on. Partitioning is also known as *fragmentation*.

2. *Replication*. Data is kept in multiple copies, usually at different sites, and usually with the intent that each copy should be identical at all times. Replicas may be complete databases, complete files within a database, or subsets of files. This latter case is a variation of horizontal partitioning, in that data may be assigned to more than one site on the basis of a value of some attribute; this is *horizontal partitioning with overlap*.

The definition of data distribution seem to imply the splitting up of a centralized database, but the combination of already separate sites is also accommodated since the rules for assignment of data to sites are also the rules for finding data.

It has been generally considered desirable that the details of data distribution be concealed from applications programmers and end-users, for the same reasons the a DBMS conceals the details of physical data storage. This concealment is referred to as *location transparency* and *replica transparency,* depending on which mode of distribution is involved. As will be demonstrated below, complete transparency is not possible in all cases; and, as will be discussed in Section 7.4, it may not be desirable in every case.

Certain constraints should be imposed on a database to ensure that semantic integrity is maintained during updates. For example, in an order-processing system the DBMS should not allow an order to reference a non-existent customer number. In a distributed environment these *referential integrity constraints* should apply across sites. However, to the extent that such constraints are part of the DBMS they are expressed in the data definition language, and consequently are limited in their scope by the scope of the conceptual schema.

Every distributed database involves data distributed by some combination of the modes described above. What makes the difference between various classes of distributed database management system is the way in which control, and the data structures (schemas, catalogues, directories, etc.) that represent control, are distributed and shared. Looking back at the Larson model in Figure 7-3, there are five cases, each corresponding to a degree of control distribution:

1. *Zeroth-Degree Distribution*. The distribution schema is positioned below the internal schema. This is a single database in which data is distributed over multiple sites, but control is completely centralized.

This configuration supports integrity constraints across sites, and provides location and replica transparency, but gives no local autonomy. Its principal use is to provide reliability through redundancy.

This case is of minimal interest, and we shall not discuss it further.

2. *First-Degree Distribution*. The distribution schema is positioned below the conceptual schema, so that there are multiple internal schemas. This is a tightly coupled database in which data is distributed, and database administration is partly decentralized—the database administrator at each site controls storage allocation and possibly physical storage organization through the local internal schema—but application administration is centralized in the conceptual schema which governs database content and logical organization.

This configuration supports integrity constraints across sites, supports location and replica transparency, and provides a limited degree of local autonomy. Its principal use is to provide the benefits of local processing—such as reduced communications cost and reduced resource contention—to users of a database which can be distributed but still needs central control over the integrity constraints.

3. *Second-Degree Distribution*. The distribution schema is positioned below the external schema, so that one external schema may be applied to multiple conceptual schemas. This is a *federation* of databases in which data and control (both database and application administration) are distributed, and which is defined by an external schema that interfaces directly with the conceptual schemas of the constituent databases.

Federations of this sort could be called "closed" federations in that new members may not be added to an existing federation (or new federations formed) unless the conceptual data elements to be included support a simple unified view from the joint external schema. (By "simple" we mean that the data definition language of the joint external schema is equivalent in expressive power to that of the individual external schemas; i.e., it does not express more complex constructs.) In order to do this, the definitions of shared data be closely coordinated. For example, if CUSTOMER_NUMBER is a data element of the joint external schema, and an equivalent element exists in each conceptual schema, then the conceptual elements should be of the same data type and the data instances should be unique across the sites. The application administrators must have agreed on the data type for CUSTOMER_NUMBER, and on a partitioning of the domain, say, 1 to M in site A and M+1 to N in site B. This cooperation is, of course, limited to those things that are to be shared across the databases.

It is not impossible for the constituent databases of this case to be of different data models; however, since the data definition language of the joint external schema can only express relationships that map transparently to the constituent, only trivial federations could be constructed across data models.

This configuration does not support integrity constraints across sites (except by the external cooperation described above), though it can provide location and replica transparency and considerable local autonomy.

4. *Third-Degree Distribution*. The distribution schema is placed between the base-level external schema and a higher-level external schema which serves as a federation schema. This is a federation of databases in which both data and control are distributed, and which is defined through

data definition information *exported* from the external schemas of the constituent databases to the federation schema.

Federations of this sort could be called "open" federations in that new members may be taken into an existing federation (or new federations formed) as long as the data definition information exported from the external level of the constituents can be accommodated in the unifying view of the federation schema. By "unifying" we mean that the data definition language of the federation schema *either* will have more expressive power than that of the constituents, which implies transparent mediation between disparate constructs to provide a unified view, or will let some of the differences show through in a manageable and consistent way. For example, if CUSTOMER_NUMBER is a data element of two databases to be federated, but is of a different data-type and overlapping domain, then the federation DDL must provide either:

- The ability to define CUSTOMER NUMBER as a synthetic element, translated to a common data-type, and mapped to a coherent domain; or
- Make the distribution apparent, by defining two elements, SITEA_CUSTOMER_NUMBER and SITEB_CUSTOMER_NUMBER, reflecting the underlying differences.

This configuration does not support integrity constraints across sites, can provide some location and replica transparency, and does provide almost complete local autonomy.

5. *Fourth-Degree Distribution*. The distribution schema is positioned below the application itself. This is a set of independent, possibly centralized, databases connected only through the application. The distribution schema is not part of the DBMS, but an information resource to the application. It serves as a directory—analogous to the white and yellow pages of the telephone directory—providing access guidance but not the access interface. That interface would have to be provided by the application, and would be defined by individual agreement between database administrators, or through standard protocols such as the ANSI Z39.50 information retrieval protocol.

This configuration does not support integrity constraints across sites, and does not provide any location or replica transparency at all; it does provide the maximum local autonomy.

These five classes are identified by Larson, though he uses different names for them.

7.2 DATA MODELS

A data model defines a set of consistent rules by which data are structured, and a set of operations that may be performed on the data.

Data modelling has been, and continues to be an important activity in database research. It grew out of attempts to formalize the data structures used in early DBMSs, and now provides the basis of much of database theory. There is an obvious value in having one universal data model as a foundation for both unified theory and standardized practice; however, to date no model exhibits the generality and power required for universal applicability.

It has been suggested that the predicate calculus can serve as a complete and integrated data model [Kowalski 1974]. While anything that can be defined consistently can be defined using the predicate calculus, it is not particularly convenient. Tsichritzis [Tsichritzis and Lochovsky 1982] has pointed out that the predicate calculus is not limited in its application to this problem by its cryptic notation—which can be replaced by more readable variants—but by its lack of strict typing. The predicate calculus is a loosely typed model that mixes data and categories of data in a formal and uniform descriptive environment without *ad hoc* restrictions, whereas strictly typed data models separate them and make strong assumptions to limit the kinds of categories of data and the kind of relationships between them that may be specified. These artificial restrictions are necessary so that the data models can be understood by human beings who are using them to formalize their view of some portion of the "real world." In short, the predicate calculus is too powerful to provide a basis for standard practice.

As far as a practical view of data is concerned, there is ample reason to believe that no single model is appropriate to all users and/or problems. As far as a theoretical view is concerned, we expect models to have consistent logical properties; therefore, to the extent that a universal model is possible, it may take on different but equivalent forms. For example, the set theoretic and graph theoretic views of data structures can be shown to be interconvertable, though they appear dissimilar and lead to very different representations of any given data structure. We can expect, therefore, that open systems will routinely involve a variety of data models for the foreseeable future.

7.2.1 Data Models For Analysis And Design

The Entity-Relationship model [Chen 1976] has gained wide acceptance as a systems analysis and design tool. It is based on entities, with attributes, whose instances are recorded in entity sets, and interentity

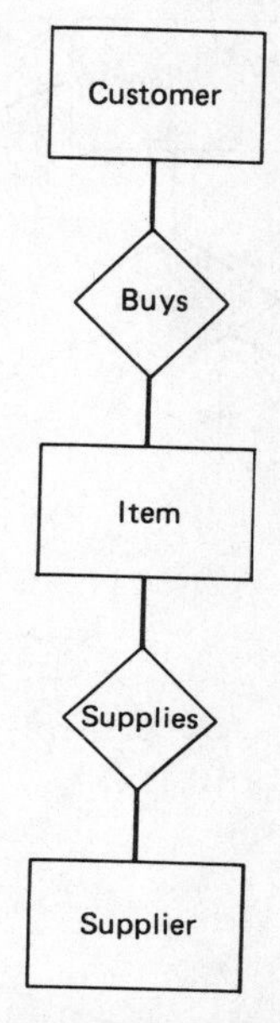

Figure 7.4. Simplified ER diagram of example enterprise.

relationships, with attributes, whose instances are recorded in relationship sets. One of the expressed aims of the ER model is specification of an *enterprise schema,* that is, a description of the data representing the activities of an enterprise of interest and the world in which it operates.

Let us consider a sample enterprise, a wholesaler that stocks various items purchased from various suppliers and that sells them on demand to various customers. The simplest possible ER model for this enterprise, as shown in Figure 7-4, consists of three entity sets—*customer, item,* and *supplier*—and two relationship sets—*buys* and *supplies*. Each of the entity sets has certain attributes that map to specific value sets (domains). Membership in an entity set is determined by a predicate; i.e., a given instance may be evaluated to determine if it is a member of a given set. Membership in entity sets is not necessarily mutually disjoint: in our example a customer could also be a supplier. Relationships may also have attributes (which also map to value sets). The entity sets that participate in a relationship set are indicated by the arcs (lines) that connect them to the relationship set; the maximum cardinality of the entity set in the relationship is given by a label on the arc, with a letter indicating ''many.''

This view of our sample enterprise is obviously oversimplified because it does not account for the stages in which a relationship such as *buys* is realized. In the world of this enterprise, sales begin as orders, which become shipments, which are invoiced to become receivables, which are

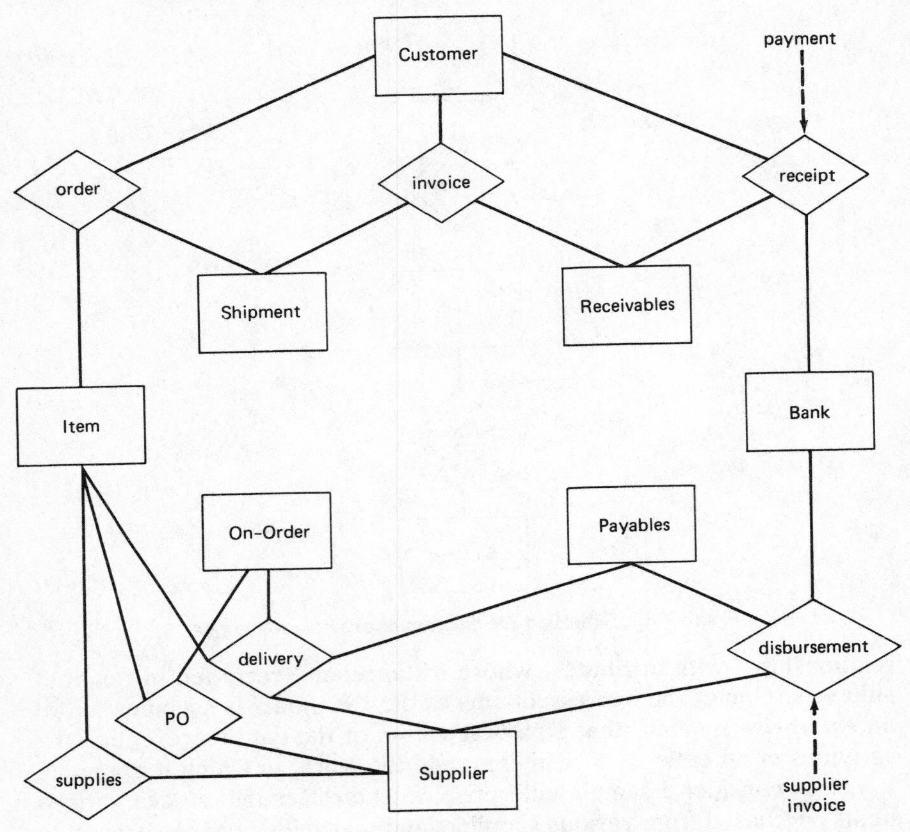

Figure 7.5. Complete ER diagram of example enterprise.

eventually paid for and the receipts deposited. A similar elaboration applies to the relationship *supplies*. A somewhat more complete representation of the example is shown in Figure 7-5. We have explicitly specified some new categories of entities and relationships, which had been previously subsumed in broader categories, in order that the model might represent more precisely the nature of the enterprise and the important restrictions that apply. This also illustrates the use of *n-ary* relationships, as in *supplies* which relates many items to many suppliers. Such relationships are useful in understanding an enterprise, but are difficult to implement in a DBMS.

As a design tool, the ER model is sufficiently powerful to model the static structures of an enterprise. It does not have any indwelling mechanism to specify dynamic rules of behavior. In our example there is no way to show directly the relationship between item quantity, sale order activity, and reorder (purchase) activity. A notation would be needed to reveal that for each item there is some lower limit to the quantity-on-hand that triggers a reorder of some minimum quantity, possibly based on recent sales volume. Petri nets, mentioned as a design tool in Chapter 3, can be used to express such rules. Kung [Kung 1984] has demonstrated that the ER model can be combined with Petri nets as a more powerful modelling tool; however, the housekeeping and consistency checking required in such a model if it were used on "real" problems would be complex enough to require automated support.

Simplicity is usually one of the stated goals of a data model, descriptive power another. There is some tension between these two goals, analogous to the tension between loosely typed and strongly typed data models. Tsichritzis and Lochovsky [1982, p. 274] have observed:

> "In any theoretical foundation, the results are only as good as the assumptions underlying the theoretical foundations. In data modeling sometimes the assumptions are rather strong which makes the theory (often called database theory) interesting but not readily applicable. However, database theory usually provides a deeper understanding of data models, database schema and their properties. Thus it should be treated as a tool for understanding and not necessarily a tool for design."

The database designer, therefore, has to work with two sets of tools: one theoretical, the other practical. Each enhances understanding only to the extent that it is simple, and correctness only to the extent that it is powerful (and in particular to the extent that its consistency checking is automated); and neither maps directly into the other. The theoretical set is one or more design-level conceptual data models, such as the ER model, with ancillary support techniques, and the practical set is the data model or models that forms the framework of the DBMS to be used.

7.2.2 Data Models for Centralized Database Management

As the framework of a DBMS, a data model is intended to provide a suitable level of abstraction for the description and manipulation of data; that is, it must obscure the details of implementation, while providing a simple and powerful interface to the designer and programmer.

Three different data models have been used as the framework of DBMSs that have achieved any widespread production use. They are:

1. Hierarchical
2. Network
3. Relational

The descriptions of these data models that follow are only superficial overviews, intended to identify the properties relevant to a discussion of distributed databases. For more complete descriptions the reader should consult a standard database reference, such as [Date 1983] or [Weiderhold 1983].

The Hierarchical Data Model The hierarchical data model, used in IBM's IMS, and in SYSTEM 2000, represents data as a tree-structure of record types. Our example enterprise might be represented as two trees: one rooted in *customer,* the other in *supplier.* The customer record might have as descendants *order, invoice,* and *receipt; order* might have *detail* as a descendant, and *detail* might have *item* as a descendant. (In this case, the paper forms used in manual processing to record transactions have been mapped to the record types.) The instances of each record type would be connected to their parent instances by pointers. The *supplier* tree would be similarly structured, with one of its leaves being *item.* In order to avoid duplicate copies of item instances, one in each tree, one would be physically stored as a pointer segment.

To a larger extent, the hierarchical model grew out of the bill-of-materials problem. Though it is no longer considered of great interest, it is widely used because of the large installed IMS base.

The Network Data Model The network model, used in HP's IMAGE DBMS and in CODASYL-conforming DBMSs such as TOTAL and IDMS, represents data as a network of record types. Some record types are designated as *masters* and other as *details,* with the details of a given type linked to a particular master type together forming a set. Our example enterprise might be represented with *customer, item,* and *supplier* as masters. *Customer* might have sets of *order,* sets of *invoice,* and sets of *receipt* linked to it; each order might have a set of *detail* linked to it *and* to *item,* forming a network of two explicitly joined trees. Similar networks would be formed between *supplier* and *item* through a *supplies* set, and so forth. The instances of members of a set are joined by pointers, which are, at least in concept, visible to the application programmer.

The network model grew out of attempts to provide a standard DBMS data model with greater generality and flexibility than the hierarchical model. It, too, is no longer considered of great interest, though it also is widely used because of the large installed base of CODASYL-conforming DBMSs.

Both the network and the hierarchical model present the programmer with the problem of *navigation* in the database, that is, the problem of traversing the network or hierarchy to perform record-by-record processing.

The Relational Data Model The relational model, first proposed by Codd [1970] grew out of a theoretical, mathematically based view of data. It represents data in simple, flat files called *relations* or *tables*. Each table represents one entity type, with the attributes of that entity type being represented by named columns. Each attribute name must be unique; each attribute must be atomic, and may assume values only from a defined domain. Each instance of the entity type forms a row of the table, and is called a *tuple*. Each tuple must be unique, as determined by the values of some subset (or all) of its attributes. The tuples in a relation are a set in the mathematical sense: an unordered collection of unique individuals.

Our example enterprise might be represented by several tables, as illustrated in part by this hypothetical DDL (which will be used in some processing examples below):

```
customers: TABLE;
  c#: integer (customer number);
  Cname: string (customer name);
  Caddr: string (customer address);
  Cphone: string (customer phone number).
items: TABLE;
  i#: integer (item number);
  iname: string (item name);
  itype: string (item type);
  iprice: integer (item price);
  icost: integer (item cost);
  iqty: integer (quantity on hand).
orders: TABLE;
  o#: integer (order number);
  oi#: integer (item number ordered);
  oiqty: integer (quantity ordered);
  oi$: integer (extended price);
  oc#: integer (ordering customer number);
  odate:
    oday: integer (day);
    omonth: integer (month);
    oyear: integer (year).
```

The relational model includes a set of operations on tables, such as:

- *Select,* which retrieves tuples from a table based on the value of some attribute or attributes.
- *Project,* which produces a relation containing only some of the attributes of a given relation (and consequently fewer tuples because of the uniqueness constraint).
- *Join,* which combines tuples from two or more tables, often using common values of an attribute in each table on which to base the

combination. (For example, *customers* might be joined to *orders* on the common values of *c#* and *oc#* to produce a table of customer orders.)

These operations are based on mathematical operations on sets, and subscribe to the algebraic principle of *closure,* which requires the result of an operation on two objects to be an object of the same type: operations on tables produce tables.

The level of abstraction and the regularity of the relational model frees the programmer from navigational problems and any direct interaction with underlying file structures. It also allows the programmer to deal with data in sets, either native to the database or constructed with relational operations as needed, instead of in individual records. (As powerful as this is, it is also very expensive in time and computing resources, so most relational implementations provide facilities for ordered access to tuples, and for tuple-at-a-time processing.)

A number of relational DBMSs are now available, including INGRESS, ORACLE, and IBM's DB2, for a wide range of machines, from microcomputers to mainframes. The relational DDL/DML SQL is already defined by an ANSI standard, and is on its way to becoming an ISO standard. It appears that the relational model will become the standard for most data management applications.

It is worth noting that the relational model as it stands has a number of limitations, particularly in dealing with temporal structures and dependencies, complex objects (such as those used in CAD/CAM applications and multi-media databases), and bibliographic structures and text. Various research projects are exploring possible extensions to the relational model. Some relational DBMSs, such as Battelle's DM, provide facilities to deal with some of these problems by making *ad hoc* extensions to the relational model, such as structured attributes and keyword indexing, and by providing access to the underlying file structures to be used where needed.

7.2.3 Data Models in Distributed Database Management

Homogeneous Databases If all of the participating sites in a distributed database use the same data model, they are said to be homogeneous. For first degree distribution (or lesser), this is true by necessity; for second degree (or greater) this may or may not be the case. Even when second degree (or greater) distributed database is homogeneous with respect to data model, there are other compatibility issues. These are discussed in Section 7.3 below.

The management of distribution resides in the distribution schema, located at varying levels according to the orders of distribution defined in Section 7.1 above. If location transparency is fully implemented, then the conceptual schema does not need to reveal any information about distribution to the programmer or end-user. Thus we would expect that the data model itself would have no influence on distribution, and the models used for centralized database management would apply.

Consider our example enterprise database. It might be distributed over a number of processing sites, say a head office, several branch sales offices, and several warehouses. The head office site might store all records pertaining to suppliers and all purchases; each warehouse might store records for its own inventory, and the sales offices might store the records relating to their customers. Each site therefore holds a fragment of the entire database: a vertical distribution of entity types between head office, sales office and warehouse; and a horizontal partitioning of entities of a given type within a given site category. (The horizontal partitioning of *items* amongst the warehouse sites would include overlap if different warehouses stocked the same items, therefore replication of items on a record-by-record basis would also be involved.)

Though a hierarchical or a network data model could support this distribution, the fundamental structures of these models cause side-effects in a distributed environment that lead to serious inefficiencies. Under both models the links (i.e., pointers) that form the structure must cross sites. If the example were implemented under the network model (where each *order detail* is linked into two chains, one for the *order,* and one for the *item*), then the *order* chains might well be confined to one site, but the *item* chains would wander around from site to site in the order that details were added. Navigation through this chain would be unpredictably slow and require many inter-site message exchanges. Similar difficulties arise under the hierarchical model, where the same logical segment is embedded in more than one logical hierarchy only one physical copy is kept and a pointer chain is constructed to link it to the physical segments of the various structures that use it.

On the other hand, the relational model does not introduce special inefficiencies since the operations on relations use sets of tuples at a time, and can therefore be partitioned site-by-site. For example, a query to find what customers had purchased a certain class of item does not have to wander through item-order-customer chains one link at a time, but can extract item numbers meeting the selection criteria at the warehouse sites, then dispatch those numbers as a set to each sales office site to be processed against the order tuples to select a set of customer numbers. The number of inter-site messages is significantly lower, and the duration of the operation is proportional to the size of the sets not to the unknown order of a chain.

Several homogeneous DDBMSs have been developed. TANDEM's ENCOMPASS/ENSCRIBE supports first degree distribution, as does

IBM's CICS/ISC. These are production systems, and were reviewed briefly in Chapter 2 above. IBM's experimental System R* and CCA's experimental SDD-1 demonstrated second degree distribution, and were the source of major contributions to the research literature.

Heterogeneous Databases In second degree distribution (or greater), the problem arises of combining several databases built under different data models. These are known as heterogeneous databases. The DDL and the DML presented to the user of a federation schema must appear to be complete and consistent; any differences in the underlying DDLs and DMLs must be made transparent. The same problem is encountered in attempting to make one type of database appear like another type. Since relational DMLs (e.g., SQL) are gaining popularity, and many organizations have large existing nonrelational databases, there is considerable interest in developing a relational interface to network or hierarchical DBMSs.

Work to date on this problem has had only limited success. IBM is rumored to have attempted to develop a relational interface to the hierarchical IMS, but abandoned the effort in favour of a separate relational database, DB2, with facilities to load data easily from IMS but not vice-versa.

Relational interfaces to network databases have had somewhat more success, with several vendors claiming to provide a relational interface to a centralized network DBMSs (e.g., SEED). The network model seems somewhat more tractable under a relational view, for reasons that are not completely clear. P.M.D. Gray offers one clue to this:

> "In some sense the set pointers [in the network model] behave as 'precomputed' join links, and we can regard many Codasyl databases as relational databases where the match operation for the join has been precomputed by joining the relevant records together in a chain." [Gray 1985]

The major experimental distributed DBMSs that have attempted to combine heterogeneous DBMSs have involved relational and network models: MULTIBASE by Computer Corporation Of America [Smith et al. 1981], and DDTS by Honeywell [Ceri 1984].

MULTIBASE uses the functional data model and the DAPLEX DDL/DML [Shipman 1981], which can express both relational and network understructures. Its architecture would correspond to second or third degree distribution in our descriptive framework. At each site a local DAPLEX schema is derived from the local conceptual schema of each participating DB, and a global DAPLEX schema is created to manage the global view. At each site, at run time, the local database interface (part of MULTIBASE) translates global DAPLEX queries into queries understood by the local DBMS. The system does not handle updates.

DDTS (distributed database testbed system) uses a data model based on the ER model, called the Entity-Category-Relationship model. A direct relational equivalent is also used. The architecture is based on the 5-level schema shown in Fig 7-1, and corresponds to third-degree distribution in our reference architecture. The participating DBs may be network or relational. DDTS supports both distributed query processing and distributed updates with full transaction management.

The problem of providing a uniform view of heterogeneous databases has as yet only limited *ad hoc* solutions. Application level protocols, such as the ANSI Z39.50 Information Retrieval Protocol, which represent the fourth degree of distribution in our reference architecture, are perhaps a more immediately practical way of combining heterogeneous databases. Such protocols provide a standard definition of message formats and interaction sequences between cooperating agents (i.e., programs in the participating sites) for database access and update. To date the protocols defined are essentially data interchange services operating in a pairwise fashion. No unified view of a group of participating databases is provided, nor are query optimization or transaction management services possible.

7.3 CONCEPTUAL DESIGN ISSUES

In a centralized database, the conceptual design issues center on the organization of the data in terms of the data model and the applications. These problems are often referred to as *schema design*.

In first and second degree distributed databases, the conceptual design problem is not greatly complicated by the distribution. Since location (and replica) transparency is generally a desirable feature, the schema of any given database does not need to reveal any information about distribution except in the form of attributes that are used to control horizontal partitioning. In our example enterprise, the customer relation is partitioned over the branch sites so a branch code might be added as an attribute of this relation. Similarly a warehouse code might be added to the item relation. Since the distribution reflects the organization of the enterprise, these attributes have a meaning in the real world and are not an artificial construct. However, the location of tuples in the partitioned order relation is fully dependent on the location of the corresponding customer tuple. Does the order relation require a branch code? While it does not appear necessary, if it is not included then it is impossible for users to control the scope of queries that refer to orders rather than customers. Therefore, the location transparency is slightly compromised, in that an artifact of the distribution enters into the definition of the data where it is not necessary.

In third degree distributed databases, the conceptual design problem becomes one of providing a unified view of separately defined and administered databases, of obscuring differences that may already exist. The objective here is to create a composite federation schema that provides the application *logical* data independence as well as physical data independence; i.e., a level of symbolic description that is independent of the logical definitions and constraints present in the underlying data model of each of the participating databases. Ceri and Pelagatti [1984] have identified four major classes of problems that arise in attempting to build composite schemata:

1. *Name conflicts,* in which the same entity or attribute is identified by different names in different databases or different entities or attributes have the same name
2. *Scale conflicts,* in which the same attributes in different databases are expressed in different units of measure (e.g., feet and meters)
3. *Structural conflicts,* in which the same fact is represented as an independent entity in one database and as an attribute of some other entity in another database
4. *Different levels of abstraction,* in which facts are more finely divided in one database than another

These problems are the reason that truly independent databases cannot be handled as second degree distributed. The federation schema of third degree distribution is required to provide the conceptual synthesis. In many cases this synthesis implies mediating processes activated by using the federation schema; for example, conversion between values expressed in different units of measure might be accomplished by routines invoked through the federation schema.

Fourth degree distribution does not present a conceptual design problem in the same sense as second or third degree. The fourth degree distribution schema does not attempt to provide a unified view or coherent synthesis of the available databases, but data and protocol definitions that must be used by the application itself.

7.4 QUERY PROCESSING

Query processing is the most frequently performed database activity, since it is used for *ad hoc* retrieval, for report generation, and for retrieving records for update. It must be efficient. The problem of query processing is not simply that of translating and executing the retrieval statements of the data manipulation language, but rather that of determining the optimum sequence of steps in performing a given query. Optimality is defined in terms least time to completion and/or least resource cost, which usually means least number of physical IO operations.

In general, query optimization techniques involve:

- The evaluation of retrieval statements in the DML to produce a "least action" version in a standard form
- The comparison of possible retrieval strategies based on the current contents of the database (file sizes and dispositions) and the cost of various operations
- The generation of a sequence of DBMS primitive actions to perform the query at lowest cost

Such techniques are, in concept, possible for DMLs based on any data model. In addition to the deleterious side-effects the hierarchical and network data models can have on performance in a distributed environment (as discussed in Section 7.2.3 above), even in a centralized environment these data models are more difficult to optimize. They require the user to navigate through the database, so their DMLs are given to small-scale stepwise procedural operations which are not suitable to large-scale optimization. On the other hand, DMLs based on the relational model, SQL for example, are given to large-scale nonprocedural operations which are suitable to optimization. This arises from the relational model's mathematical basis, and is, of course, one of its strengths. Most of the work that has been done on query optimization—both in centralized and in distributed systems—has been done in the context of the relational data model and relational DMLs.

One of the tensions between conflicting database requirements shows up in query processing. The binding-time problem, that is, the choice between precompiled queries and dynamically interpreted queries, is a reflection of the conflict between efficiency and flexibility. The overhead of analyzing and optimizing a query is incurred every time a query is processed if interpretation is used, but only once if the queries are precompiled. Since a DBMS should support both repetitive queries and *ad hoc* queries, it would seem that the solution is to provide both techniques; however, that means developing and maintaining two separate routines. In addition, the existence of precompiled queries can limit the flexibility of the database by making it difficult to make simple housekeeping changes, such as altering a view or dropping an index, where there are precompiled queries that could be rendered inoperable thereby. This problem is more acute in distributed situations where data relocation possibilities and local autonomy issues increase the need for flexibility, and at the same time make it harder to keep track of the precompiled queries which can be held remotely.

7.4.1 Query Processing in Centralized Databases

In order to illustrate the kinds of optimization that can be performed in a centralized database consider the following query against our sample wholesaler database described in Section 7.2.3:

```
Find the names and phone numbers of customers who ordered
cosmetics in June.
```

Or, restated in SQL on the relational version:

```
SELECT UNIQUE cname,cphone FROM items, orders, customers
WHERE itype=''cosmetics'' and i#=oi# and omonth=06 and
oc#=c#;
```

To flesh out the example, assume that the database contains:

- 500 customer tuples (1 per physical block)
- 5000 items (5 per physical block), of which 10% are cosmetics
- 1000 order-events/month, average 5 items/order-event, 12 months or 60,000 order tuples (10 per physical block)

The example query involves several join operations. Joins are formally defined in the relational model as a *theta selection* on a *cartesian product*. As such, more than two tables can be joined in the same query and the joins can be performed in one operation. In the example that would involve creating a set of tuples with one entry for each combination of the tuples of *item, order,* and *customer,* and the applying the selection operators to keep only those that satisfied the conditions of the WHERE statement. (The UNIQUE specified in the example represents a projection operation to be applied after the join operation, which was included only for completeness and which will not be discussed further.)

The cartesian product is never explicitly formed in the processing of a query involving a join because it is simply too costly. In the example, 150 billion tuples would be formed in the cartesian product (500 × 5,000 × 60,000), requiring 3 billion physical block reads (500 × 1,000 ×

6,000). However, one of the methods used to perform joins, the *nested loops* method, proceeds in a manner analogous to forming a cartesian product, by reading sequentially through one relation, and for each tuple in that table reading sequentially through the relation to be joined to it, and so on. The improvement in the nested loops method is that the restriction operation (i.e., the theta select) is applied at the appropriate stage to reduce the number of reads performed.

A nested loop join, with the nesting as stated in the example query (*customer* within *order* within *item*) would cost 3.251 million physical reads (1000 + (500 × (6000 + (1 × 500)))):

- Read items (5,000 tuples, 1,000 blocks) select 500 cosmetics
- For each item selected read orders (60,000 tuples, 6,000 blocks) select 1 (5,000 in June × .1 cosmetics/500 cosmetic items)
- For each order selected read customers (500).

If the nesting were *customer* within *item* within *order*, then the number of physical reads would be more than 5.256 million (6000 + (5000 × (1000 + (.1 × 500)))):

- Read orders (60,000 tuples, 6,000 blocks) select 5,000 in June
- For each order selected read items (5,000 tuples 1,000 blocks) select .1 (1 item number match × 10% cosmetics)
- For each item selected read customers (500)

The efficiency of the method is sensitive to the order of the nesting based on the size of the relations and on the expected number of selections at each stage. The performance can also often be increased if the semantics of the database are taken into account. If the fact that item numbers and customer numbers are unique is considered in the preceding example, some loops can terminate after the first match has been found, thereby reducing the number of reads in the loop by half. In the first order of nesting shown above, the customer read loop can terminate after a selection has been made, reducing the number of reads to 3.126 million; in the second order, both the item loop and the customer loop can terminate after the first selection, reducing the number of reads to 2.631 million.

An alternative method for processing joins is known as *merge-scan*. This method relies on sorting the relations on the join attribute value and performing the selections in the read and merge phases. For the example query, a merge scan might take as few as 65,400 physical reads and writes (assuming sorting is an $N \log N$ process, where N is the number of blocks in the file):

	TUPLES	BLOCKS
read items	5,000	1,000 read
select *itype*=cosmetic	500	100 write
sort selected items		500 IO
read orders	60,000	6,000 read
select *omonth*=06	5,000	500 write
sort selected orders		3,100 IO
merge selected items and orders &	5,500	600 read
select orders on *inumb* matches		500 write (max)
sort customers	5,000	42,600 IO
merge customers and selected orders &		5,500 read
select customers on *C#* match		5,000 writes (max)

This is considerably more efficient than the nested loops method in this case, and indeed in most cases where the relations are relatively large and stored without order or indexing. The merge-scan does require temporary file storage and is considerably more complex to construct; it does not usually benefit from database semantics and is often less sensitive to the order of evaluation. (In the example, performance is independent of the order of the first two reads and sorts; however, if the selections on item and order were not performed at all until the merge phase, the total number of physical IO operations would be 120,400.)

If the relations are stored in order or indexed by one or more access keys, then both selections and joins can be performed using index values. If in our example database, the relations *order, item,* and *customer* are indexed on *order#, item#,* and *customer#,* respectively, and retrieval of a record through an index takes 3 read operations, then a nested loops search, in the order specified (customer within order within item), would cost over 3 million physical IO operations (1000 + (500 × (6000 + (1 × 3)))):

- Read items (5,000 tuples, 1,000 blocks), select 500 cosmetics
- For each item selected read orders (60,000 tuples, 6,000 blocks) average select 1 (5,000 in June × 10% cosmetics/500 cosmetics)
- For each order selected read *customer* on *customer#*

On the other hand, a nested loops search in the order *customer* within *item* within *order* would cost only 23,000 physical IO operations (6000 + (5000 × (3 + (.1 × 3))) + 500):

- Read orders (60,000 tuples, 6,000 blocks) select 5,000 in June
- For each order selected read item on item# select .1 (10% cosmetics)
- For each item selected read customer on customer# from order
- Write each customer (maximum 500)

Further, if the *order* relation were indexed by *item#* and *customer#* as well as *order#,* then a nested loops search in the order *customer* within *order* within *item* would cost about 10,000 physical IO operations (1,000 + (500 × ((3+11) + (1 × 3))) + 500):

- Read items (5,000 tuples, 1,000 blocks) select 500 cosmetics
- For each *item* selected, read *order* on *item#* (3 reads to first record with that key, 1 to each of the average 11 thereafter: 60,000 order details/5,000 items) select orders in June, average 1;
- For each order selected read customer on customer#

With indexing (or relations stored in order on a join attribute) the merge-scan method will not yield any significant improvement over the nested loops method.

What the foregoing examples show is that information can be traded for phsyical IO operations. Order, of course, is information; and where it does not exist—as with unsorted, unindexed relations—it is often more efficient to create it than to proceed without it. The merge-scan method vastly outperforms the nested-loops method on unordered relations, but it is much more complex and requires its own built-in information in the form of programming. When order is maintained—as with index relations—the nested-loops method becomes more efficient. In this case the built-in information comes from the programming of the index maintenance routines, and from information saved at update time (but at a cost in processing overhead for each update). What matters most in query processing is the effective use of available information.

7.4.2 Query Processing in Distributed Databases

Query processing in a distributed environment must deal with the same problems as in a centralized one, plus the problem of the cost, both in time and dollars, of data communication. In a distributed context, the order in which actions are carried out influence not only the total number of file read and write operations, but also the amount of data and the number of messages that have to be transferred from site to site. The site at which a query originates may not store any of the relevant data, so that control of the processing may also have to be transferred.

Consider the query analyzed in the preceding section, applied to a distributed version of the sample enterprise database. Assume that the distribution is 1st degree, comprising a head office site at which *supplier* and purchase-related data is stored, a number of branch offices at which *customer* and sales-related data is stored, and a number of warehouses at which *item* data is stored.

For instance, if the query were processed using the nested loops method, even with the benefit of optimal indexing, and one site were the processing site, then a message would be generated at each iteration of the item loop to be sent to each warehouse site where it would search the *item* table and return a single item tuple. The number of messages, and consequently the message wait time would be very high. On the other hand, using the merge-scan method would entail shipping large volumes of data (entire item tables for example) between sites. To cope with this problem, a technique called a *semi-join* is used.

A semi-join involves breaking the query processing steps into a projection of the attribute or attributes of interest, shipping them to another site to perform a semi-join, and shipping the results back to the originating site to complete the join. Less data is moved, because not all tuples survive the semi-join, so this should be faster and cost less. In our example, a select order might be sent to each warehouse site to retrieve only the *o#* of *item* tuples matching the select criteria; these would be returned to the originating site which would distribute the projection of the combined result to each branch office site to be joined with the *order* tables to return only the *C#* of the tuples that result, and so on.

Compared to the centralized case, we can see that the complexity of query optimization has increased significantly with the introduction of the semi-join. There are more steps and many more possible orderings of the steps to be evaluated. System R* for example attempts to evaluate all possibilities, comparing nested loop and merge-scan methods using *shipped-whole* or *fetched-as-needed* options here applicable. (Note that cost of shipping a whole relation may not be prohibitive in a high-bandwidth network such as an Ethernet LAN.) It attempts to optimize the query processing in terms of processing and communications cost and time.

It has been shown that for the general case this optimization problem is equivalent to the *traveling salesman* problem; that is, it is computationally intractable if perfect results are expected. Nonetheless, good approximations are possible in constrained cases if sufficient information is available to the optimizer. This is true of first degree distribution, but in second and third degree distribution information about the underlying structures—especially such useful facts as the existence of secondary index files—is not necessarily present in the higher-level schema. In these case specific actions might be necessary to export such information to support query optimization. In the fourth degree, optimization is left entirely to the application.

7.5 TRANSACTION PROCESSING

The paramount issue in database management is the preservation of the integrity of the data, given that both hardware and software can, and often do, fail. As long as the database is used only for query processing, there is no problem; the stored image of the database does not change and can be restored from a backup image should it be lost. However, once updates are allowed, the stored image does change over time, and cannot be restored without loss of work if the changes are not themselves backed up. In a batch processing environment the job that made the changes can simply be rerun; in an interactive environment the changes must recorded separately so that they can be reinstalled should the primary database image be lost.

The notion of database integrity cannot be limited to individual update operations on records. Semantic integrity constraints imply that sequences of reads and writes are often necessary to effect a complete and consistent update in a logical sense; for example, an order processing application might reasonably require that only complete orders—headers, all detail lines, and totals—appear in the database. Clearly this requires a number of writes to add and update records. The concept of packaging a related group of individual update steps as a single *transaction* first appeared in the literature in Eswaran [1976] and is essential to reliable database management.

A transaction is a ''delimited sequence of database operations which together perform a logical unit of work'' [Lindsay et al. 1979]; that is, a sequence of update steps which change the content of a database from one consistent state to another. The precise nature of what constitutes a consistent state, given the limitations of current data modelling techniques, is known only to the application; therefore the scope of a transaction—the delimitation of start and finish—must be defined by the application program. The act of completing a transaction after the application has indicated that it is finished is known as a *commit;* if for any reason the transaction is not allowed to commit, it is said to *abort*.

It has also been shown that in a multi-user environment where several transactions may be in progress concurrently, possibly attempting to read or modify the same data elements, database integrity can only be ensured if the result of concurrent transactions is the same as if they had been executed one after the other in some (unspecified) order. This is known as the *serializability* constraint.

7.5.1 Transaction Processing in Centralized DB's

Transaction Logging and Recovery Transactions are particularly vulnerable to hardware and software failure. No matter what techniques are used, at the time the updates are made to the stored data a device failure and/or a software failure can leave the database in an unknown and potentially inconsistent state. For this reason, transaction processing necessarily involves mechanisms to record the progress of transactions in a log file held in stable storage (disk), and to recover from various failures. Haerder and Reuter [1983] have conducted a thorough analysis of the techniques to support transaction-oriented database recovery, and have proposed an "ACID" test of a system's quality, notably that a transaction must be:

*A*tomic	*All* changes made must be present in the database at the completion, or *none* must remain
*C*onsistent	At completion the transaction leaves the database in a logically consistent state
*I*solated	Changes made by a transaction must be invisible to other concurrent transactions (and vice versa)
*D*urable	Once committed the system must guarantee that the results of the transaction survive subsequent malfunctions

The *atomicity* property requires that the system manage the updates made individually by the transaction in such a way as to ensure that they are collectively applied or discarded. In practice, this involves applying the updates in controlled stages, in one of two ways:

1. Updates are aplied to the database as presented by the application, but only after a log record is made of the before-image (the unchanged version), sometimes called *prompt update*.
2. As updates are presented a log record is made of the after-image (the changed version), but the update is not applied to the database until commit time, sometimes called *deferred update*.

If the system fails before a transaction commits, or if it decides to abort the transaction, or if the application decides to abandon the transaction, atomicity is preserved in one of two ways:

1. In the prompt update case, the before-images from the log file are used to *roll back* or *undo* the changes made by the transaction.
2. In the deferred update case, the after-images in the log file are abandoned so that no changes are made.

If the application orders the transaction committed, in each case the log file is marked to so indicate. If the system fails after a transaction has committed, durability is ensured in one of two ways:

1. In the prompt update case, nothing is done to committed transactions.

2. In the deferred update case, the after-images are used to *roll forward* or *redo* the changes made by a committed transaction.

The reader should be aware that the foregoing description has been simplified to clarify the important principles. There are a number of other technical issues and trade-offs, for example buffer management and IO optimization, that complicate these mechanisms. Some systems employ both before-image and after-image logging to deal with special problems. (Systems using before-image logging often also provide after-image logging as a faster, sequential source of update images to supply an incremental database backup process, rather than copying changes from the database itself to a backup medium.)

Concurrency Control In an interactive environment, it is usual for many users to be performing transactions at the same time. The *isolation* property and the *consistency* property, to the extent that they are directly enforceable by the DBMS, require that accesses to the database by concurrently executing transactions be synchronized so that they do not interfere with each other. If this is not the case, a variety of problems can occur that will compromise, or appear to compromise, the consistency of the database, such as:

Lost Updates. If two concurrent transactions A and B (originating with different users in a multi-user environment) both modify record X—A reads X, B reads X, each updates X and finally commits—then the update issued by the first transaction to commit will be lost to (overwritten by) the update issued by the second to commit. If A was crediting your bank account and B was concurrently debiting it, the order in which they commit might be of considerable interest.

Inconsistent Reads—Individual Records. If two concurrent transactions A and B both access record X—A reads and updates X, B reads updated X, the A aborts—then B has seen an inconsistent database state. If A was an erroneous attempt to charge a large amount against your credit card account, and B was a inquiry to determine if your credit limit was sufficient for some purchase, a read by B of A's attempted update might also be of considerable interest.

Inconsistent Reads—Sets of Records. If two transactions A and B are using the same *set* of records X{i}—A is reading each X{i: for i = 1 to End} to total some value in X, while B is performing transactions against various individual records in X{i} which change the value being summed by A—then when A completes the total it has computed is not necessarily consistent with any database state. It is clearly not the total of X{i} at the time A started, and if B updated any record that A had already visited, then it is not the total at the time A completed or of any time in between.

The mechanisms to effect the necessary synchronization are known as *concurrency control,* and are part of the transaction discipline. In a multi-user environment, where concurrent activities are possible, the transaction discipline needs to be applied not only to sequences of actions that change the content of the database, but also to those that only read the content (as can be seen from the inconsistent read problem foregoing).

It is clear that the isolation and consistency of concurrent transactions is preserved if the concurrency control mechanisms allow overlapped execution of transactions only to the extent that the result is equivalent to a *serial* (one-at-a-time) execution sequence of the transactions.

The method most widely used to achieve serializability is *locking*. As a program reads records within a transaction, it requests that they be locked; that is, recorded by the DBMS in some global data structure as in-use. Locks may be *shared,* allowing other transactions to read but not update and indicating that the locking transaction itself has no intention to update; or they may be *exclusive,* preventing other transactions from reading and indicating the locking transaction's intent to update. In some systems the lock level must be designated at read-time, and in others locks set at read-time are always shared, and at write-time are promoted to exclusive if possible. The DBMS ensures that one transaction may not place an exclusive lock on a record locked by another, nor place a shared lock on a record exclusively locked by another, and in consequence the transaction not have access to the record until it is unlocked. Obviously the DBMS will not allow a transaction to update a record which is not exclusively locked by that transaction.

It has been shown that *two-phase locking* (2PL) guarantees a serializable schedule [Eswaran et al. 1976]. In 2PL a transaction reads and locks all of the records which it will use before it issues any updates and/or unlocks. Under a transaction discipline with the atomicity and isolation properties, the locks are held until the transaction commits or aborts.

When a transaction encounters a record to which it cannot have access because it currently belongs to another transaction, it may be made to wait until the blocking transaction has completed. In most cases this will work satisfactorily and the delay will be transparent to the application program. However, it is clearly possible for pathological situations to arise, for example:

given two concurrent transactions A and B, A exclusively locks record X, B exclusively locks record Y, A attempts to lock Y and is made to wait until B releases it, B attempts to lock X and is made to wait until A releases it.

The result is a *deadly embrace* or *deadlock* in which A is waiting for B which is waiting for A. Since the number of concurrent transactions can be greater than two, it is possible for deadlock cycles to be longer, and therefore more difficult to detect.

There are two principal techniques available for detecting and resolving deadlocks:

1. Cycle Checkers. These inspect wait lists to determine if there are loops. This checking can be performed when a transaction is being made to wait, or it can be performed periodically by an asynchronous process. When a deadlock is detected, it can be broken by choosing a victim–such as the current transaction, or the one that has done the least work—and aborting that transaction, releasing all its locked records. The aborted transaction must restart from the beginning, while those waiting for its locked records may proceed.

2. Timeouts. When a transaction is made to wait, it is assigned a time limit by the application or the DBMS. When that time has passed and the transaction has not been allowed to proceed it is aborted on the presumption that it is involved in a deadlock and all its locked records are released.

Since it is possible for an aborted transaction to restart so quickly as to immediately recreate the deadlock (especially in the case of timeouts where it is possible that two have timed out simultaneously), there should be a short, random delay, perhaps exponential in the number of retries, applied to the restart.

Concurrency control using locks can be made deadlock free by aborting the transaction attempting to lock a record already locked by another transaction, rather than waiting. This might increase the number of retries, but it eliminates any need for deadlock detection and resolution.

There are two techniques other than locking that have been applied to effect concurrency control: *timestamping*, and *optimistic* concurrency control.

Several timestamping algorithms have been proposed [Bernstein et al. 1980]. A time value (either a clock time or system-wide sequence number) is applied to each transaction. When a transaction reads or updates a record its timestamp is appended to the record. Conflicts are detected when a transaction asks to read a record that has been updated by a younger transaction, or to update a record that has been read or updated by a younger transaction. In this case, the requesting transaction may be aborted and restarted with a new timestamp. (Some variants of the method allow the requesting transaction to wait under certain conditions [Rosenkrantz et al. 1978], but these are not deadlock free.) Where locking guarantees the execution to be equivalent to *some* serial execution, timestamping guarantees it to be equivalent to the order of the timestamps.

Optimistic concurrency control [Kung and Robinson 1981] assumes that conflicts are rare, and does not check until the application program asks that the transaction be committed. The transaction's updates are withheld from the database until commit time, and are committed only if the records read by this transaction (whether or not they were updated) have not been updated by any transaction that completed since this transaction began. The DBMS must keep a record of the read and write sets of each in-process transaction, and of the write sets of transactions completed while other transactions are in-progress. If a conflict is noted, the transaction is aborted, rather than committed, and made to restart.

The reader might well wonder about the relative merits of each of these techniques. The important factors are the relative complexity of the code, and the run-time performance: practical trade-offs rather than any formal difference. Early simulation and analytical studies were inconclusive, and often contradictory; for example, some studies reported that blocking algorithms (i.e., those that wait on detection of a conflict) performed better than those that immediately abort and restart [Carey and Stonebreaker 1984], while other studies indicated exactly the opposite [Tay and Suri 1984]. A recent study [Agrawal et al. 1985] compared the throughput and response time performance of blocking, restarting and optimistic algorithms using a simulation model with realistic load and memory/CPU/IO resource constraints. It indicated that previous studies were correct, *within their assumptions about system resources*, and demonstrated that the relative performance of the algorithms varies with the load and available resources. In particular, the multiprogramming level within the DBMS (i.e., the number of transactions currently competing for resources) strongly influences the performance.

An interesting sidelight of the Agrawal study was that it showed no statistically significant difference in the relative performance of the three algorithms in a simulation of 200 terminals performing transactions on a 10,000 page database, at 4–12 pages/transaction with 1 second think time between transactions. Only when the database size was reduced to 1,000 pages did the conflict rate become large enough to reveal differences. This confirms the experience of many practitioners: in ordinary operational situations conflicts are rare, typically less than .1% of transactions wait, unless there are "hot spots" in the data access patterns. (Such hot spots are often artifacts of the DBMS or database design rather than of natural usage patterns.) Gray has proposed a simple analytical model of the probability of waiting and deadlock [Gray et al. 1981a] that supports these observations and indicates that:

- Waits and deadlocks are indeed rare, and both increase linearly with multiprogramming.
- Essentially all deadlock cycles are of length two.
- Waits increase as the second power, and deadlocks as the fourth power, of transaction size (in pages).

This model was proposed as a "straw man" to be demolished by a more careful analysis; however, it seems quite durable. Applying it to the Agrawal study, in the 10,000 record case the probability of a transaction waiting is .18, of deadlock .0013; in the 1,000 record case the probability of a wait is one, and of deadlock .13. In other words, performance differences become apparent only when the probability of any conflict has been raised to an unrealistic degree.

Given this rarity of conflicts, the most important aspect of run-time concurrency control performance is the processing overhead for each transaction, not for abort/restart. If the choice is left to the application programmer, it is not unusual for concurrency control to be ignored in order to avoid this processing overhead and the consequent throughput and response-time penalties. Nonetheless, conflicts do occur and database integrity cannot be guaranteed without concurrency control. Unless it is fully integrated into the DBMS and made as transparent as possible to the application programmer, concurrency control can be easily defeated.

The only transaction management error that should be passed back to an application is one indicating that the transaction has been aborted and must begin again with its original external input data. There are systems in which the error reported is that a requested record is unavailable, and the programmer is left to decide how to deal with the situation: wait, unlock records and restart, or just try again. Such sloppy protection can lead to program hangs and system aborts as serious as the problems arising in the absence of any concurrency control.

7.5.2 Transaction Processing for Distributed DBs

In a distributed database system, a transaction originates with a process executing an application program in a given site but may perform reads and writes on portions of the database held on other sites. A component of the database management system must manage the transaction at the originating site, and cooperating components must manage the transaction steps performed at each participating site. These components, sometimes referred to as *DBMS agents,* are linked through the communication subsystem.

A distributed database management system faces, therefore, not only more complex synchronization problems than a centralized DBMS but also more and different modes of failure. The synchronization problems are not, in themselves, substantially different from those encountered in centralized transaction processing; the major differences arise from the modes of communication failure:

- Messages may be lost in transmission.
- Messages may not arrive in the order sent due to transmission errors and consequent retransmission.
- Certain failures make it impossible to tell if it is a remote site or the connection to that site that is down.

In the latter case, it is possible for a network to be *partitioned* by a communication failure so that it is divided into two (or more) separate groups of sites. Each group comprises one or more sites that can communicate with each other (i.e., within its own group) but not with any site in another group (i.e., across the partition). Database agents may continue to operate in isolation from their peers, making decisions which could affect the integrity of the database if there are outstanding (i.e., incomplete) transactions that span the partition.

Concurrency control is complicated by the intervening communications subsystems. If a locking protocol with waiting is used, deadlock cycles are global in that they may span several sites. If a global cycle checker is used, as in System R*, the algorithm is quite complex and generates some communication traffic of its own. If timeouts are used, as in TANDEM'S ENCOMPASS, spurious timeouts can occur due to communications congestion or processing overload at remote sites. In either case, communications failures can leave old locks "hanging about" if the site managing the transaction that owns them becomes inaccessible.

In a centralized environment, when an application program signals completion of a transaction, the DBMS performs the necessary actions to commit the transaction, or to abort it if a conflict is discovered. In a distributed environment, the possibility of communications failure during this phase is particularly critical, since it is possible that one site might carry out transaction commit while another decides to abort. This situation requires at least a *two-phase commit* protocol (2PC) to ensure that the atomicity of transactions is preserved across many sites.

In a 2PC, when the application issues a commit order the controlling site sends a *prepare* order to all of the sites involved in the transaction. Each participating site must acknowledge the prepare with either an *accept* or a *reject;* if it accepts, a site may not thereafter unilaterally abort its part of the transaction. If all participants responds with accept, the controlling site will issue a *commit* order to all; if any site responds with reject, or does not respond within some specified time, the controlling site must issue an *abort* order to all participants.

The management of transactions on replicated data is particularly troublesome. Replicas can be managed with a master-slave technique in which reads are made against a local (slave) copy and updates are made against the master and propagated at a later time to the slave copies. This technique is employed in TANDEM'S EMPACT system, which is a second degree distributed system build on ENCOMPASS for internal corporate use in production management. It does not provide replica transparency in the proper sense, because updates to the master copy can be made only by its owner, and because applications reading slave copies may be reading old information. Where the slave copies are primarily for read-only use, it is nevertheless quite acceptable.

Replica transparency can be provided only with algorithms that ensure coordinated updates within the transaction. The simplest of these is the Read-Any-Write-All technique. If updates to the replicated data are rare with respect to reads, then it is an acceptable practice, otherwise it is much more costly than a centralized approach. It also suffers from communication failures in that updates are not possible if any copy is inaccessible. Algorithms that read-any-write-most (but not all) have been devised, but are very complex and have not been shown to be practicable in any real environment.

Analogously with query processing, transaction management becomes more complex with each degree of distribution, and depends heavily on shared information and mutually compatible algorithms. The techniques used in centralized systems have been extended as described above, and used in production first degree systems and experimental second degree homogeneous systems. With heterogeneous second degree and third degree systems, it is not necessarily the case that the underlying local DBMS will use the same algorithms or exhibit complete transaction integrity with respect to concurrency control and commit protocols. In these cases, specific mediating agents must be constructed within each site to create an overall transaction management mechanism.

A general case of such mediating agents is being defined as part of the OSI Common Application Service Elements (CASE). The CASE concurrency control, commit, and recovery protocol now being developed as an ISO standard allows for a tree of processes spanning multiple systems and connected through the session layer to provide locking and commit services in a relatively general way; however, these must be coupled to the underlying local data management services. As vendors develop their own implementations of this standard, the possibilities for heterogeneous second degree, third degree, and fourth degree distribution will increase.

7.6 PHYSICAL DESIGN ISSUES

In a centralized database system, physical design issues are largely concerned with the problem of improving performance while containing storage cost within limits.

The per-byte cost of disk storage hardware has declined steadily at a rate of about 20% per year. The capacity of individual devices has increased by several orders of magnitude in that time. A second scale level of device, the 5.25'' winchester, appeared in the late 1970s, and is showing similar price/capacity trends. Access times and data transfer rates have also improved. Nonetheless, performance/storage cost trade-offs continue to be important, largely due to increases in interactive systems and in the volume of data held on disk for on-line use. The example query discussed in Section 7.4.1 illustrated the importance of index files in improving access performance; however, extensive indexing can consume as much or more space than the data itself.

While disk access time is usually the largest component of total access time, in-memory processing, especially searching, sorting, and matching operations, can also contribute significantly. In a multi-user situation, managing and optimizing such processing is a major concern. In a contentious environment, with a database subject to increasing size and/or processing load, the total cost of database operations is non-linear, rising more rapidly with increasing size/use. The net benefit, expressed in the same units as cost, is also nonlinear, flattening with increased size/use, due largely to the concomitant increase in response time (or decrease in throughput). Weiderhold has suggested that this phenomenon creates a ''safe zone'', as shown in Figure 7-6, inside of which database operation is cost/effective, but not outside [Weiderhold 1983]. This is a physical reason for distributing a database: after a certain size and/or usage intensity it may be more cost-effective to split it up and escape the excessive cost penalty incurred in centralization.

This situation is also the impetus for the development of specialized hardware, usually known as *database machines,* to improve the performance of database systems. The essential reasoning is that database operations, especially searching and matching, can benefit from parallel-

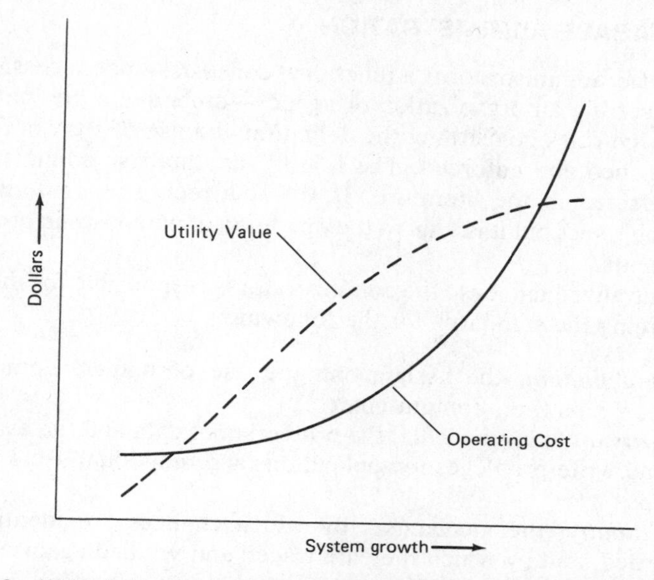

Figure 7.6. Weiderhold model of database cost/benefit under growth in size and use.

ism and special-purpose hardware. For example, it is obvious that if a search of an index block (i.e., a key comparison) can be made as data is transferred from disk, rather than after it has been moved into a memory buffer, the operation will take less time and fewer resources. Even moving the database processing to a separate but otherwise standard computer may provide significant improvements since the optimization within the database machine can be tailored to database operations, unconstrained by general OS and application considerations.

A database machine in a local area network configuration is often called a *database server* to indicate both its specialized role and its shared nature.

Physical design issues arise for distributed databases only in systems of first or second order where a unified database is being fragmented or there is otherwise a central administration authority over the entire database. The issues still center on the cost/performance trade-off; however, to the problems of the centralized environment are added the cost of communications and the cost of replication. Communications costs, both dollars and time, can be reduced by replicating data where queries are significantly more frequent than updates. On the other hand, since updates must be propagated to all copies, at a cost in communications services and time, replication is a costly strategy where updates are more frequent.

7.7 DATABASE ADMINISTRATION

The database administrator is a functional construct—not necessarily one person, possibly an organization or agency—embodying the mechanism by which the rules concerning the definition and use of a given database are established and enforced. The role of the database administrator is often neglected in the literature. It is not a technically interesting or challenging topic, but it is vital to the operation of databases in production environments.

In a centralized database the administrator is responsible for the setting and enforcing the standards for the following:

- *Data definition,* the assignment and use of names, domains and relationships (i.e., content control)
- *Security and access rights,* the protection of data and the assignment of read/write privileges to applications and individual users and user groups
- *Auditability,* the mechanism by which changes are identified and recorded, and by which they are traced and verified against supporting external information
- *Operational integrity,* physical backup and recovery mechanisms, control over program testing and periodic integrity checks
- *Operational performance,* resource allocation and mechanisms by which access to the database can be optimized to meet users' needs (includes performance monitoring and analysis)

The database administrator is responsible for the state of the database, both form and content, in exactly the same way in which the chief financial officer of a company is responsible for the state of its books. If this responsibility is not brought together under one person or organizational unit, then there will be no accountability and no coordination of various database policies and practices. An important aspect of the database administrator's duties is the management of trade-offs. For example, it is often necessary to trade away some security or integrity control for increased efficiency. If this decision can be made by those not responsible for the security or integrity, then expediency will almost always overrule prudence. It is the database administrator's task to control this.

In a distributed environment, as local autonomy increases it becomes increasingly difficult to bring database administration under a single authority. In first degree distribution the administrative authority can remain centralized. If the sites will respond to the administrator's operational performance demands then there is no difference from the centralized case. In second degree distribution, the sites may have more operational autonomy and the participating databases may have independent administrators; therefore, the administration of the distributed database formed by the participants is a matter of cooperation and compromise. This becomes more pronounced in third degree distribution; in fourth degree administration, it largely vanishes to be replaced by adherence to standards and pairwise agreements between users and database operators. It is interesting to note that one of the few evaluations of database administration in a distributed environment to appear in the literature [Walker 1982] defines the primary task of the administrator as communication between users, organization management, system operators, and applications maintenance staff.

One of the reasons that more distributed databases are not in place is that the tools to support administration in distributed environments do not exist. The EMPACT manufacturing control system developed by TANDEM Computers for its own use in a distributed environment, a second degree distributed system in our terminology, was one of the earliest examples of a production system of this type. After a few months of operation its developers observed:

> "Operationally, however, the need for a centrally located network management [i.e., database administration] function has become apparent; events that require the cooperation of all the different sites, such as the addition of a new site to the network [i.e., to the distributed database], are very difficult to manage manually in a distributed environment." [Norman and Anderton 1983]

The management of a distributed database requires a "database" to manage the metadata or distribution schema that defines and controls the distributed database. This database may itself need to be distributed; in particular, replicas of it or parts of it may be needed at every operational site, though one copy might reasonably serve as a master to coordinate updates from the various participants in the database administration process. It is this problem of describing and coordinating database definitions that is unique to the distributed database environment.

7.8 DEFINITION AND DESCRIPTION OF DISTRIBUTED DATABASES

The problem of managing the data structures that define a distributed database was raised in the previous section. These structures themselves form a kind of database which itself may be distributed. They are the physical realization of the schema layers described in the ANSI/SPARC and Larson architectural models, are known by a variety of names, including: schema file, database definition file, database directory, data dictionary, or database catalog.

In a centralized DBMS, these data structures are generally obscured from the application programmer or end-user, and the data they contain made available where necessary by a utility interface. In the early stages of database management these structures, and indeed the metadata, were considered as extrinsic to the data management problem. The initial aim was *physical data independent,* so that application programs and databases might be changed independently of each other. It did not then seem reasonable to think of programs that might "plug in" to multiple, different databases at run time. *Logical data independence* requires that a program, or a mediating agent acting on its behalf, be able to adapt different conceptual representations of data to a unifying higher-level conceptual representation. This means that metadata should itself be accessible, at least for query, as though it were in a database.

Several DBMSs built on the relational model have implemented their database descriptions in the relational form embedded in the database they describe. Such DBMSs are said to be *homeomorphic*. IBM'S DB2 follows this practice with a catalog containing several distinguished indwelling relations (tables): SYSTABLES a relation naming and describing the relations (tables) in the database, SYSCOLUMNS naming and describing the attributes (columns), and so on [Date 1984]. SQL queries may be performed on the catalog.

In a distributed environment of second or third degree, provision and management of shared metadata is a significant problem. The distribution schema is not subordinate to the control of a single database administrator, and therefore needs to be maintained as a quasi-independent, shared entity; i.e., as a database. In a distributed environment of the fourth degree, the distribution schema may be a more general structure describing a group of available databases that support certain interface protocols or provide a certain class of service. In this form the descriptive structure is not so much a schema that provides a definitive unifying view, but a directory that provides hints and access mechanisms.

Associated with distributed databases is a family of descriptive data structures of varying degrees of authority and complexity serving the following functions:

- *Dictionary services,* name to description translation
- *Name registration,* ensuring the uniqueness (within context) of a name
- *Authority services,* associating variant forms of a name with a standard form
- *Thesaurus services,* establishing hierarchies and networks of classes

- *"White page" directories,* name to access path translation
- *"Yellow page" directories,* classification to name/access path translation

There is as yet no standard for such data, though several attempts are under way. In particular, both ANSI and ISO are considering a proposed standard for Information Resource Directory Systems (IRDS). The proposed standard is in fact a definition of a general purpose database to manage information that describes other databases, processing agents, and protocols. It originated with the National Bureau of Standards in the U.S., and in the ANSI version is expressed in the ER data model to be implementation free. The ISO version has been restated in the relational model using the proposed standard SQL; however, the requirements of the proposal are such that they cannot be expressed in SQL as it stands, principally because of SQL's lack of referential integrity constraints. This in itself illuminates some of the problems in data management for open systems.

The directory problems discussed in Chapter 4 are also relevant in this context.

7.9 COMMUNICATIONS WITHIN A DISTRIBUTED DATABASE SYSTEM

In a centralized database management system the individual software modules can be designed to interface by flow of control, that is, through procedure calls. The communication system that links interactive users with the DBMS is entirely external to the DBMS.

On the other hand, in a distributed DBMS at least some of the modules must be designed to communicate with each other via explicit interfaces to the host system's communications subsystem. The DDBMS must be "aware" of the separation of its parts, and of the nature of the communications paths between them. There are several architectural "styles" that can be applied to structure of communications within a DDBMS, with varying implications for efficiency and standardization.

The nature of distributed transaction management is such that the DDBMS designer can never assume reliable communications. Not only can the loss of a message compromise the integrity of a transaction, in some cases, a significant delay in the transmission of a message can also have similar effect. For example, even if the communications subsystem guarantees the eventual transmission of a message in the event of a communications link failure, the DDBMS cannot rely on this, say, to commit a prepared transaction because hardware failure conditions and/or timeouts at the remote site could intervene. Therefore, the

DDBMS designer will rely on message passing between peers (i.e., DBMS agents of equal authority) with end-to-end acknowledgment.

This style is in the spirit of the ISO Open System Interconnect model (see Chapter 6 for a discussion of this model). However, if the peer-to-peer exchange is implemented as a session within the ISO/OSI definition, as in System R* and the OSI CASE CCR protocol, significant processing overheads may be incurred. A session is a connection between exactly two peers. Since distributed transactions may involve more than two sites, the transaction manager at each site must manage a session linking it with each other site in the network, or at least participating in the current transaction. If a transaction manager process is created for every concurrent transaction, a great deal of extra bookkeeping is generated, not to mention the response time penalty paid if the session has to be established for each transaction.

An alternative is to provide a message-passing capability to the DDBMS that behaves like a *datagram* service, accepting independent, explicitly addressed messages for delivery to remote DDBMS agents. This approach shows some similar to the Bachman architectural model of a DDBMS in which the database management layers are embedded in the ISO/OSI model below the session layer, thereby having privileged access to lower-level communications services.

Broadcast techniques have been considered for use in DDBMSs. A DDBMS agent at one site creates a message which is copied to DDBMS agents at *all* sites concurrently (but *not* simultaneously). Agents who can act on the message do so and acknowledge; others discard the message. This style of communications may have some advantages for query processing in high-traffic configurations including replicas in relatively many copies. A query can be sent to all possible respondents, and the first site that is able to process the query responds. It may also be useful for propagating updates to multiple copy replicas and for transaction commit messages in certain environments. In general, broadcast techniques work efficiently only when supported by physical broadcast media, such as CSMA/CD links using coaxial cable or radio frequency transmission.

Chapter 8

Conclusions: Future Developments in Open Systems

It should be clear at this point that the problems to be faced in the building of open systems are more complex than we have been accustomed to thinking. The process of implementing genuinely useful open systems environments even in North America and Europe, both of which seem to be ripe markets, will be one of long, slow evolution. Most of us tend to conceive open systems mainly in terms of communications: the ISO Open Systems Interconnection model and so forth. Although undeniably important, the communications issues related to open systems interconnection represent only one aspect of a larger problem. Indeed, purely technological issues form only part of the problem.

In this book we have attempted a more comprehensive view of open systems than has been customary. There have been two main areas of consideration. One of these is of course technological: how rapid changes in the technology itself will advance, or in some cases retard, the realization of open systems. The other is essentially historical and political: we have an existing base of computer systems, and we'll have to continue to live with them. Moreover, we'll have to *agree* on the means of incorporating these, as well as a myriad of new systems and devices, into a coherent, integrated open systems environment.

The computer field has changed radically over the past 25 years. It used to be that the customer went to the vendor with a problem, and the vendor told him how to solve it. The buyer didn't pretend to understand the solution; he simply accepted it, much as he would a pill from a physician. The vendor was thereby invested with a doctor's mysterious power. The advantage of this arrangement was of course that the vendor could comfortably limit his solution to the wares he has available.

Today's buyers are more sophisticated; they know both the problems and the basic means of solving them. Much of the mystery has been dissolved by familiarity and education. Computers are even sold through catalogs, and the customer can call a Zenith number to specify the bits and pieces he wants.

What the buyers of today want are open systems. Vendors are faced with an increasingly strident demand from their user population for all the benefits of interconnection, and at all levels. The organizations that have most to gain from an open system environment are the biggest ones. To the Boeings and General Motors of the world the computer industry must listen whether it wants to or not. Because a good part of what the market wants is beyond the capability of current technology, their demands have created the problems.

Big customers cannot only demand solutions, they can specify particular types of solutions, and these often become *de facto* industry standards. These are easier to manage than formal, negotiated standards, and more popular because they don't require so much negotiation and work, but they are not always for the best because they are derived only from one usage. AT&T's UNIX is one of the happier instances: while it may not be the ideal operating system, programmers like it, and it has proven itself both flexible and portable and capable of improvement.

Sometimes *de facto* standards are the result merely of the bullying of one powerful organization, which may or may not recognize even its own best interests. Still another way of promoting such standards is the "open specification." For example, in order to gain acceptance for its Multibus II standard, Intel has published the full specification. For a nominal sum any company can gain full Intel support for its implementations of that specification.

On the one hand, *de facto* standards can facilitate interconnection, and on the other they exist, and must be considered in the design of open systems.

The demand for interconnection has also caused a rapid proliferation in formal standards and guidelines. Standards are agreed procedures, defined in great detail, and formulated by cooperative organizations. Guidelines are looser definitions of general approaches to problems. Each discipline now has its own set of standards: for example, communications, database management systems, and now software engineering. The influence of national standards groupings such as ANSI and IEEE has grown steadily with the level of connectivity within North America. And since the open system environment is ultimately international, the International Standards Organization has assumed great prominence in the computer world over the past few years.

The burdensome cost of research and development is one of the main reasons both government and industry promote standards, for standards are designed to ensure that R&D investment is long rather than short range. On the other hand, since technology is changing at a dizzying pace, standards must be formulated with great care and foresight: they must take into account the introduction of devices and procedures not yet even imagined.

The premature introduction of an inappropriate standard can cripple new technology. Failure to agree on a single standard where one is plainly required can cause great inconvenience and expense in the long run. One example is the British habit of driving on the left, while everybody else drives on the right-hand side of the road. Although it would plainly be convenient for the British to change their ways, at this point the conversion would be very disruptive. Another is the use of 220-volt electrical power in Europe and 110-volt in North America; it would be better for all if a single standard were adopted, but the changeover would be prohibitively expensive. The worldwide telephone system is one case in which a single, appropriate standard has been acceptable to all; the result is that we can all talk to one another.

No one vendor or group of vendors can think of supplying solutions to all the problems. Even in single technological areas enormous resources are required, sometimes so great that not even individual national governments can bear the burden alone. During the 1980s in particular a great number of research consortia have grown up. Some, like MCC in the United States, are cooperative ventures involving a number of independent companies, intended to distribute the costs of basic research. So severe is the problem that governments, recognizing that economic survival is at stake, have been willing to relax antitrust legislation to permit the cooperation and sharing of resources by companies which are otherwise in competition with one another.

Other cooperative ventures are sponsored by the governments themselves: the ALVEY project in Great Britain, and the Fifth Generation Computer Project in Japan are notable instances. Japan's Ministry of International Trade and Industry (MITI) has sponsored an impressive range of projects designed to bring that country into the front ranks of the computer industry.

Finally, there are now several instances of serious international cooperation in basic research. One good example is the ESPRIT project sponsored by the European Economic Community. Participating in ESPRIT are the governments of all ECC member countries, companies from most of these countries, and indeed three large American firms.

The aims of these consortia and joint ventures, both national and international, are to share the costs of basic research and to avoid the duplication of effort in such research. Since the results are to be shared by all participants, there will be a common ground in the products offered by each. This is in effect still another form of standardization. In the development of an open systems environment the issue becomes one of communication not only among governments and businesses, but also among consortia, to ensure at least a general compatibility.

Although the problems, technological, historical, and finally political, are daunting, the resources brought to bear on these problems are increasingly impressive. The result will surely be the gradual institution of a world-wide open systems environment. "Gradual," however, is the key word: not this year or next, but perhaps ten to twenty years down the line.

8.1 BASIC TECHNOLOGICAL TRENDS

We can at least predict with some confidence some of the technological developments of the next ten years or so, and in the light of these we can foresee some of the directions in distributed and open systems.

8.1.1 Communications

One of the most exciting prospects lies in optical fiber technology, which offers the possibility of almost unlimited bandwidth in data communications. Optical fibers already carry data in point-to-point communications pipes at rates in the order of hundreds of gigabits per second. Bell Laboratories and several Japanese companies have been leapfrogging in an ongoing race to provide the highest transmission rates over the longest distances. In 1985 Bell Labs managed to transmit at a rate of 4 billion bits per second over a link of 117 km—without the use of repeaters [Bell 1986].

The product of the data rate and the distance has been doubling every year since 1975; we can increase the present figure by five orders of magnitude (i.e., 16 years of doubling) before we reach the limits imposed by physics. In the use of a megawatt laser 10 photons (quanta of electromagnetic energy) are needed to detect one bit of information. A single strand of fiber should be able to support a data bandwidth of 10^{15} bits per second. That could be divided (for example) into one 100 megabit-per-second channel for *each* of 10 million users.

Communications facilities based on light-wave technology are being installed all over North America. A number of vendors offer local area networks based on optical fiber. In this application, however, the technology is not yet mature: more research is needed into the efficient tapping of light pipes. As soon as this problem (the loss of light in tapping the main pipe) is resolved, fiber-optic LANs will become commonplace.

This planet is already laced with computer communications networks of all sorts and sizes: local area, wide area, metropolitan, packet-switched, circuit-switched, satellite, packet radio, and cellular radio. Most of these are not immediately compatible with one another. The devices employed by the end-users of such networks are equally diverse and incompatible: telephone, personal computers, large host machines, PBX switches, alarm systems, video systems, FAX systems, and so forth. The problem of incompatibility is plainly aggravated in an open system environment.

What is needed in an open system is a standard digital communication service which can connect any device with any other, whether the two be across the room or around the world from one another. The ISDN is designed to provide an integrated solution to this problem. In the ISDN a customer interface (a plug in the wall) is defined, whereby the user can attach his device and gain access to a worldwide network. We are not likely to see much definition or implementation of ISDN components until the beginning of the next decade. However, a timetable with clearly defined milestones has been established. What that means is that we are approaching a time when an astonishing level of connectivity among systems and devices will be possible. Connectivity at a very high level will be necessary if we are to gain the full benefits of an open system environment.

8.1.2 Processors

In the realm of processors perhaps the most important development is the proliferation of 32-bit microprocessors in personal computers and workstations. These machines, small and inexpensive but powerful, are spearheading the drive toward an open system environment. At the other end of the computer spectrum, supercomputers and parallel machine architectures are emerging to increase the processing power we can apply to any single problem.

Both these fields are developing quickly, and creating a demand for ready and easy communication, which is in turn propelling research and development in open systems. Architectures for systems which can be tightly coupled, loosely coupled, and even hierarchically structured, are beginning to appear.

8.1.3 Software

Distributed and interconnected systems with enormous computational power will appear over the next ten years. Unless careful attention is paid to the user interfaces to such systems, users will be confused, and the open system environment will therefore not provide its promised benefits. At the very least we must provide the user with languages that allow him to take advantage of an open, distributed architecture to write application code quickly and to modify and maintain application packages such as distributed database management systems.

The complexity of a distributed system should be transparent to the user. He should interface to a system-wide operating system which demands only a single logon, with a network-wide name and password (provided through local, regional, national, and international network directories). The interface must provide access to file servers, database servers, automatic backup (through transparent global data directories), processing servers, mail servers, application packages, education and help facilities, and a myriad of other services. The open system itself can take advantage of expert systems facilities in providing all these services to the end user.

High reliability and fault tolerance will be ensured by extensive redundancy within the distributed system. The system should be self-repairing, and even self-organizing, as conditions and demands change.

Apart from the business-oriented applications and facilities mentioned above, there will be an equally extensive list of consumer-oriented services. One product that spans business and personal needs is the provision of easy access to the many commercial information banks. Indeed, it is not unlikely that users will be confronted with so many attractive services that they will find it quite impossible to use them all.

8.2 PRINCIPAL PARTICIPANTS IN OPEN SYSTEMS RESEARCH

Much of the research pertinent to open systems is connected, directly or indirectly, to the field of Artificial Intelligence (AI). During the 1970s, AI was a ''pure'' research activity, the principal centers being Carnegie Mellon University, M.I.T., Stanford University, and the University of Edinburgh [Scown 1985].

By the 1980s it had become apparent that AI had commercial and military potential; thus, existing large computer vendors such as DEC, Honeywell, IBM, and UNISYS, multinational corporations such as Goodyear Aerospace, Rockwell International, and Schlumberger (a multi-billion dollar oil exploration advisory firm) became significantly involved. Schlumberger alone now has AI groups in France, Connecticut, Texas, and California.

The various countries differ in their concerns and in the ultimate purpose of their research. Here are some of the more notable cooperative research programs.

1. The *Japanese* have established a Fifth Generation Technology project [Lindamood 1984, Moto-Oka 1984]. Their aim is to meet the social needs they anticipate will emerge in the next decade. They also expect that exporting AI technology will improve their world economic position. Japan's goal is to build a full-fledged fifth-generation computer system and make it available for commercial computer systems. The project has been completely funded by the Japanese government in the initial three year stage. The key money to inaugurate ICOT (Japan's institute for new-generation computer technology) was donated by a

consortium of eight manufacturers, which also provides the funds to run ICOT. The member companies are obliged to give ICOT their complete understanding, cooperation, and support. ICOT is unusual in Japan in that it is a separate, independent, neutral organization that has been established to carry out a research project. The customary Japanese approach is to have each of the participating research institutions and companies conduct work on its own. The ICOT research center plans to approach the overall goal of complete fifth-generation system by pursuing two intermediate hardware projects of top priority: a parallel inference machine and a knowledge-based machine. ICOT's major software efforts will be oriented toward two software systems: a problem-solving and inference system for processing problems and knowledge-based management system for accumulating and managing knowledge.

2. The *United States* is primarily interested in maintaining its own world leadership in information technology. The United States is also concerned with the application of AI computing technology to national security, to build up its strategic strength with respect to USSR and the communist-block nations. The US goal is to build up a greater common pool of knowledge about individual systems, both for the military and also for the use of private companies in further developing products in the usually competitive free market. United States has its roots in AI research for 20 years. The quantity of research is considerable, but it has been scattered among some 40 universities and 30 major corporations. That piecemeal approach is changing, however, and engineering and computer science resources, both corporate and academic, are coming together. There is also a growing realization that cooperative research between otherwise competing commercial companies need not threaten market share, but rather is necessary for strength in highly competitive international marketplace.

The Microelectronics and Computer Technology Corp (MCC) in Austin, Texas, is a for-profit research company formed by 21 otherwise competing corporations [Lineback 1985]. The research corporation is focusing on microelectronics packaging, computer aided design and manufacturing, software engineering, and computer architecture. Two other formed research groups, the semiconductor research cooperative (SRC) in Research Triangle Park, North Carolina, and the Microelectronics Center of North Carolina (MNC) are really research brokers that funnel money from a number of sources primarily to universities.

3. The countries of the *European Economic Community* (EEC) banded together in 1984 in the ESPRIT program (European Strategic Program for

Research in Information Technology) [ESPRIT 1984, 1985]. The goal of the program is to promote European cooperation in the development of information technology. A secondary aim is to promote European standards for the next generation of computing machinery. $1.5 billion is to be distributed to universities and industrial research organizations, to support projects in microelectronics, robotics, software engineering, information processing, office automation, and computer aided manufacturing. Each of these projects involves cooperation between industrial partners from at least two countries as well as one university. Three American-based firms, ITT, DEC, and IBM, are participating in ESPRIT projects.

4. The ALVEY program is the *United Kingdom's* fifth generation project [ALVEY 1984, Oakley 1985]. Its aim is to develop the tools and methods necessary for the production of high-quality, cost-effective hardware and software: specifically, the technology needed to support a new generation of computing and communication systems. The ALVEY Committee coordinates the efforts of government, industry, and universities. Principal research areas are software engineering, VLSI, expert knowledge-based systems, and intelligent man-machine interfaces.

Applicants to ALVEY are expected to form research consortia, carrying out research at their own facilities rather than at a central location (compare the Japanese Fifth Generation project, or MCC). An example is a consortium which is producing integrated software engineering tools. It is composed of six industrial and academic members, under the leadership of SDL, a software house. One ALVEY hardware project is the development of a new microprocessor chip, called the "transputer," for parallel processing applications.

5. In 1984 the *West German* Ministry for Reseach and Technology initiated a program which brings together industry and academic researchers in research areas such as computer-aided design, computer architecture, and knowledge engineering.

6. The *Soviet Union* has a five-year plan to develop its own fifth-generation computer. The Commission for Computer Engineering at the Moscow Academy of Sciences has set goals for the plan in the areas of VLSI, microprocessors, parallel and multiprocessor computer architectures, intelligent databases, and formal logic as a basis for computer operation. The countries of the Council for Mutual Economic Assistance (CMEA) have been asked to contribute to the Soviet program.

8.3 A WORKING CONSORTIUM: MICROELECTRONICS AND COMPUTER TECHNOLOGY CORPORATION

We have selected MCC from among the many possible examples of the new cooperative research organizations for several reasons. First, it is a large company with a comprehensive program: its work covers most of the fields cultivated by the other major national and international programs. Second, since MCC is sponsored by most of the major American computer manufacturers, it must surely represent some important part of the future of the industry. Third, the company exists to make a profit; that condition imposes a certain practicality. Finally, the MCC project schedules runs neatly parallel to a general timetable for the development of open systems.

8.3.1 Programs at MCC

The main areas of research and development at MCC are [Fischetti 1983, 1986, Limeback 1985]:

1. New packages for semiconductors, to increase the density and decrease the size of systems containing ICs.
2. Increased efficiency in the design and development of software by large teams.
3. Increased efficiency in the design of ICs through CAD.
4. Faster computation through parallel processing machines.
5. Artificial intelligence and knowledge-based systems in computers that can "reason."
6. Large, complex databases to store the information required in knowledge-based systems.
7. Better user interfaces for computers, to permit users to work more efficiently and effectively.

8.3.2 Competing with Other Co-ops

This summary of MCC activities is based on an interview by Fischetti [1986] of key MCC personnel.

MCC was initiated at a time when there was a great emphasis in the U.S.A. on cooperative research. Companies in the semiconductor and computer industries were finding that they simply could not afford the growing cost of basic research; the solution was to share it with other companies. About the time of MCC's founding several other cooperative research ventures were formed, for example: the Semiconductor Research Corp. (SRC), in Research Triangle Park, North Carolina, and the

Microelectronics and Information Sciences Center (MISC), in Minneapolis. These groups operate as consortia, channeling funds from government and corporate sponsors to laboratories and universities engaged in research. It was hoped that by sharing their results they could both bolster American industry and upgrade research at American universities. Another aim was to coordinate the basic research work of participating companies in order to avoid useless and expensive duplication.

The participants in MCC might well have chosen to join other research consortia, but chose instead to form their own. Some duplication may have resulted. There remains, for example, some question among software researchers whether there is an overlap between MCC's software program and the research of both the Software Engineering Institute at Carnegie-Mellon University in Pittsburgh and the new Software Productivity Consortium in Washington, D.C.

8.3.3 The Basic MCC Research Programs

Semiconductor Packaging Improvements in chip and circuit-board packaging are the aim of MCC's shortest-term technical program. It was originally scheduled to last six years, and is the most aggressive program in terms of deliverables. Much of the work on IC packaging is intended to turn the experimental tape-automated-bonding (TAB) method into a robust production-floor process. TAB embodies several advances in chip packaging. Most of the techniques in use today involve bonding the chip within a closed, hard-plastic package. Package leads are then soldered onto the printed circuit boards.

In TAB the chip is bonded to the surface of a tape resembling 35 mm camera film. Tiny metal connections are deposited between special pads on the chip and leads on the tape. The chip and its leads are then sliced from the tape and bonded directly to a circuit board. As a result chips can be placed much closer together on the boards: some 35 to 40% of the board area can be covered by TAB chips, as compared with an average of 1 to 5% for conventional printed-circuit boards containing chips in conventional surface-mounted packages.

The new approach permits the manufacture of much denser circuits, because the leads can be placed much closer together. In production chips, the external leads are 100 mils apart, although a few chips are made with 50 mils spacing. Chip-leads on the MCC TAB line are between 7 and 12 mils apart; by 1987 they will be down to 5 mils. Chips with 328 leads have been fabricated at the packaging laboratory in the same space needed to provide only 48 to 64 conventional leads. The payoff of TAB is that denser chips with many more leads can be placed much closer together, on boards with denser inter-chip circuitry. The ultimate goal of

Table 8.1. Milestones in Research at MCC.

1. Packaging
   - 1983–86 Develop automated bonding for factory-floor and thin-film processes
   - 1986–88 Perfect both thin-film printed-circuit boards and cooling technology
   - 1988–90 Refine design to achieve manufacturing speed and reliability

2. Software
   - 1985–86 Staffing and strategy
   - 1986–87 Research software management
   - 1987–88 Draft strawman of Leonardo
   - 1988–92 Refine and build Leonardo prototype

3. VLSI/CAD
   - 1983–86 Develop hardware accelerators, and redo mathematics
   - 1986–88 Develop CAD for 1,000,000-transistor chip
   - 1988–89 Develop CAD for 10,000,000-transistor chip
   - 1990–92 Develop ultimate CAD system for multiple designers

4. Parallel Processing
   - 1983–85 Identify applications for parallelism
   - 1985–87 Design parallel architecture models and languages
   - 1987–90 Build and evaluate proof-of-concept machines
   - 1990–94 Build prototype architectures

5. AI/KBS
   - 1983–87 Develop knowledge-based system; derive tests for KBS
   - 1987–88 Test KBS, integrating parallel processing, AI, database, and human factors

6. Database Systems
   - 1983–86 Define models of advanced database systems
   - 1986–89 Develop tests for evaluating databases
   - 1989–94 Build prototype database machines using parallel processing, logic languages, and VLSI circuits.

7. Human Factors
   - 1983–86 Experiment with interface technologies; model human users
   - 1986–88 Build intelligent user-interface management system
   - 1988–94 Build various prototype interfaces; improve management system.

SOURCE: Fischetti Mark A. "A Review of Progress at MCC" IEEE Spectrum Vol. 23, No. 3, March 1986 p.80.

the Semiconductor Packaging program at MCC is to package the circuitry of a supercomputer in a portable system like a workstation.

Software Researchers in MCC's Software Technology project have dubbed it "Leonardo," in honor of the man program director Laszlo Belady calls the last "polyhistor," an individual with mastery over the entire range of mankind's accumulated knowledge. The name has been turned into an acronym: Low-cost Exploration Offered by the Network Approach to Requirements and Design Optimization. (Leonardo can be considered "low-cost" only in the light of the expensive problems it addresses.)

Exploration is the operative word. The essential work of a designer in approaching a large systems problem is to explore a wealth of alternative solutions, select the most likely, and then test its implications for the system through prototyping, simulation, and analysis. This exploration is an iterative, creative, time-consuming process, and one for which computers offer very little assistance.

As a designer explores alternative solutions, the computer system providing assistance might point out earlier, similar problems and the solutions adopted. The system might also store the designer's decision-making process, for review when the designer returns later to the problem, or when someone else takes it up. Although some current systems aimed at assisting the designer store the design itself, they do not store the rationale behind it. This type of assistance demands the use of some elements of expert systems; creating an expert system, however, is not the goal of Leonardo. The project's primary task is to identify the activities that take place during the work of a design team, and then to seek ways of facilitating these tasks.

According to Belady, he expects Leonardo both to confirm and to meet the need for communications among the members of a design team. Such a team should therefore include specialists who concentrate on various aspects of the system: performance, communications, maintainability, reliability, and so on. All these aspects should influence the design at an early stage.

Leonardo also addresses the problem of graphic defiction of design concepts. Current graphic approaches—mainly dataflow diagrams—offer only marginal help here. Leonardo will develop a rich, two-dimensional design and specification language. Animation of the language's symbols will help designers synchronize the many interrelated processes in the design.

Among the interim deliverables generated by Leonardo thus far is an algorithm to solve the so-called *n*-party interaction problems.

All products stemming from Leonardo will be written in Ada [Kull 1985].

VLSI-CAD All design tasks at MCC will be configured with 81 multiply interactive LISP workstations that communicate with specialized high-throughput computers, or "accelerators," which MCC will build.

A key task is the development of a common user interface in which a particular function (such as "move a wire") will be achieved by the same command on all workstations. Most of today's CAD systems, by contrast, are simply stand-alone workstations linked together in a piecemeal fashion.

At MCC CAD tools are being developed for tasks ranging from process simulation to VLSI layout and system synthesis. Layout programs are geared to the "floor-plan" approach, in which the designer picks functional circuit blocks from a library of blocks, and then concentrates on the best way of interconnecting them. The floor-plan approach is seen as an element of the fourth generation in design evolution (the first three being gate arrays, standard cells, and logic arrays), and is just beginning to receive acceptance in industry.

At the same time, researchers are reworking the basic mathematics that generate design algorithms, circuit models, and circuit simulations. Later in the program, knowledge-based systems will be used to take design-for-test rules and impose them automatically on a circuit design.

To date (1986) MCC researchers have used the LISP language to create a module editor in the C programming language, a computer program that allows designers to lay out circuitry graphically. The program has 117,000 lines of code and runs on a UNIX system.

Parallel Processing The development of computers that process data in parallel rather than serially—and thus at much higher speeds—poses both hardware and software problems. The few parallel processors built thus far can execute only particular functions, and use only limited forms of parallelism to solve primarily numerical problems. The underlying principles of parallel processing have not yet been developed to a degree that might permit a designer to use these techniques in a wide range of applications.

Researchers at MCC have examined a number of programming languages that have potential in the area of parallel execution semantics. Selected applications have been written in a parallel dialect of LISP called MultiLISP, a functional language with graphic-reduction semantics developed in England, and in SISAL (for "Streams and Iteration in a Single Associated Language"), a data flow-oriented language developed at the Lawrence Livermore National Laboratory, Digital Equipment Corporation, and the University of Manchester in England.

Recently MCC has begun developing advanced languages, evaluating various parallel execution models for the languages, and designing physical computer structures to implement these models. By the end of 1986 the Parallel Processing Group will initiate the first of a series of proof-of-concept machines, in which different parallel architectures will be tested. The goal of this group is to produce high-speed parallel processor architectures that use symbolic language and reduce development and execution time by a factor of 10 to 100.

Artificial Intelligence and Knowledge-Based Systems The program in "AI-KBS" is surely the most adventurous of MCC's efforts. Its primary goal is to develop a computer that can make decisions on the basis of common-sense reasoning as well as facts, and which can solve problems in spite of uncertain or missing facts.

To this point AI has been most visible in the context of "expert" or "knowledge-based" systems: computer programs that give the same answers to questions that a human expert would. Examples are the DARPA program that identifies safe cross-country pathways for tanks on the basis of weather reports and satellite imagery. As advanced as these expert systems are, however, they can solve only clearly defined problems by following clearly defined rules dictated by human beings.

A skeleton common-sense knowledge base is to be built in 1986 by the AI-KBS group. It is to include a broad but shallow base of the knowledge that human beings consider common sense. Included will be knowledge about the knowledge base itself, permitting the system to "know" when a question is beyond its current range of understanding.

If "thinking" machines are to be useful they must produce results in a matter of seconds, not minutes or hours. MCC researchers have already developed a special computer language that will allow the imbedding of artificial intelligence techniques at the fundamental, machine-language level of a computer.

Database Systems Another key to knowledge-based systems, and to many other computer applications, is the database. This is a massive store of knowledge. Even modern bulk-storage methods using high-speed disks are slow relative to the speed at which a computer computes, and databases are growing to the extent that entering all the data may take months. The MCC Database program aims to get more information in and out faster. To this end, database machines will be built, and each machine will rely on both parallel processing and VLSI circuitry. MCC intends to build database machines several orders of magnitude more efficient than those available today.

Two different types of database languages will be developed. The first—accounting for 75% of the program's effort—will be a "logic database." In such a database data is entered and retrieved through logical (i.e., Prolog-like) assertions. The second will be an "object-oriented" database. It will make use of LISP (symbolic) processing to give the programmer a maximum of control over computation.

Although logic programming optimized for a small database that fits into the computer's main memory is under development in Japan, no one has ever built a compiler for large databases residing on disk. The MCC logic database is to implement a new language called LDL, or Logic Data

Language, essentially an upgraded version of Prolog. LDL could become a general-purpose programming language, used to make conventional databases more powerful.

Human Factors In all three of the MCC computer programs, and in the software and CAD programs, emphasis is placed on improving the efficiency with which the programmer, designer, data entrant, and end-user interact with the computer. All these programs therefore rely on the Human Factors group.

A wide range of techniques are being explored, including voice-recognition, natural language processing, and three-dimensional graphics. Linguists are helping to develop more powerful natural-language techniques, concentrating on multi-sentence dialogs. They are cooperating with AI researchers and systems designers to create a "user's assistant:" some combination of techniques which prompts the user, accepts various inputs, and answers questions. It will advise and coach users to adapt to each technique. Computer design engineers are to incorporate speech recognition and image-handling in the User's Assistant.

The Human Factors group is working its way beyond icons—the graphic symbols used on terminals to represent functions and objects. Icons slow a user's progress, and are becoming too numerous to be readily distinguishable from one another. Rather than icons the group hopes to use objects in realistic setting and in abstract diagrams, with three-dimensional graphics, to lead a user through an instructional sequence.

By the end of 1988 MCC hopes to demonstrate an intelligent user-interface management system, with general software building blocks and tools to customize these. Once this system has been built it will be used in the building of advanced interfaces.

8.4 A RECENT OPEN SYSTEMS SURVEY

In July of 1986 the Department of Communications of the Government of Canada [Canada 1986] issued the results of a survey on the future impact of open systems—mentioned briefly in Chapter 6. It seems appropriate to conclude with a discussion of this survey because it reflects the current state of opinion (in Canada at least) on many of the issues discussed in this book. It also reinforces certain of the conclusions we have drawn.

There were 39 respondents. Among the vendors of computing equipment were IBM, DEC, Philips, Honeywell, Northern Telecom, as well as a number of smaller firms such as Geac. All three of the major Canadian common carriers were represented: Telecom Canada (Bell), CNCP Telecommunications, and Teleglobe. Both the National Library of Canada and the Canadian Institute for Scientific and Technical Information (the National Science Library) submitted responses, as did several major university libraries, public libraries, and research institutions, as well as the Canadian Association of Research Libraries. Four provincial governments participated in the study; the government sector was also represented by the RCMP and by Hydro Quebec. Finally, among the respondents were the principal Canadian interest groups for industries such as banking and manufacturing, and several consultants in computing and communications.

The primary concern of the survey was to determine the potential impact of open systems (specifically of systems on the OSI model) on the marketability and competitiveness of Canadian products in the computer and communications fields. All the respondents agree that OSI will have a highly significant impact six to twelve years from now. While such unanimity is encouraging, a more interesting result is that only 91% believe that this impact will become apparent in the near term (one to five years). In parallel, all agree that OSI will contribute to a more universal accessibility of information systems and services, but only 92% believe that it will provide users with a greater choice in hardware and software suppliers.

To illustrate the import of these results we examine four responses. Two are from major computer vendors, IBM and DEC. Note that these companies, although they are subsidiaries of American firms, are Canadian companies: the opinions are not necessarily those of their parent companies, and are from a distinctly Canadian point of view. The other are major users: Telecom Canada, which controls most of the Canadian telephone system, and the Canadian Bankers Association. It is worth noting that the Canadian banking system differs from the American in that it is dominated by a few very large institutions, each of which has offices from coast to coast. In the context of open systems, the bankers are especially interesting because of their traditional reputation for secretiveness.

DEC feels that the impact of open systems will be apparent by 1987. IBM, on the other hand, foresees no dramatic effects in the near future; the demand for open systems will be in the longer term. The Canadian Bankers Association say that bankers will be ready for open systems within three years; the first significant applications will be point-of-sale networks and home banking. The common carrier believes that open systems will be with us within three years, and asserts that Telecom Canada will be ready for them.

DEC and IBM plan to implement the X.400 message-handling standard, and is already beta-testing a commercial system embodying part of that standard. IBM claims that while they are aware of a growing interest in open systems, none of their customers have made firm plans to install such systems. They do not plan to introduce products until a firm demand for them is evident. The bankers are working on a formal requirement for point-of-sale and home banking applications; the document will be completed within three years. They express disappointment that none of the major vendors or carriers have shown much interest in OSI-based systems and services. Telecom Canada, on the other hand, claim that in the near future they will be conducting commercial trials of OSI-based Teletex and message-handling services.

DEC believes that OSI will give users a greater flexibility in their choice of equipment, because they will no longer be locked into proprietary network standards. IBM says this will be only one of many considerations in selecting hardware and software, and its influence will depend greatly on the customer's need for inter-system communications. The position of the Canadian Bankers Association on this issue is that if every vendor offers OSI, then there will be no significant changes; if, however, only some offer OSI products, these vendors will have the upper hand. Telecom Canada states simply that those vendors offering OSI will have a distinct advantage.

The first conclusion to be drawn from these responses is that IBM is more skeptical than other organizations about the impact of OSI. The question is, are they skeptical because they already control a majority of the market, and see no need to share it with others, or because they believe that open systems will be much more difficult to realize than the others seem to expect. Elsewhere in their reply they are aware of real technical problems to be overcome.

A second conclusion is that communication among the buyers and sellers isn't quite what it should be. There appears to be real confusion in the customers' minds as to what the vendors are up to, and the vendors themselves aren't fully aware of what their clients want.

A third conclusion is that there is plainly a need for open systems. So clear is the requirement that there is a tendency, particularly on the part of users, to overlook technical problems. OSI (and with it the ISDN) has a somewhat magical aura.

And finally, it's going to happen; it's just a matter of when.

Bibliography

Acker, R. D., and Seaman, P. H., "Modeling Distributed Processing Across Multiple CICS/VSSites," *IBM Systems Journal,* Vol. 21, No. 4, December 1982, pp. 471–489.

ADA, "Preliminary Ada Reference Manual," *ACM SIGPLAN Notices,* Vol. 14, No. 6, June 1979.

ADA, "Reference Manual for the Ada Programming Language," ANSI/MILSTD-1815A, U.S. Department of Defense, 1980.

Adiba, Michel E., et al., "POLYPHEME: An Experience in Distributed Database System Design and Implementation," *Distributed Databases,* (Delobel, C., and Litwin, W., editors), North-Holland Publishing Company, 1980.

Adiba, Michel E., Andrade, J. M., Fernandez, F., and Toan, N. G. Nguygen, "An Overview of the POLYPHEME Distributed Databases Management System," *Information Processing 80* (Lavington, S. H., editor), North-Holland Publishing Company, 1980, pp. 475–479.

Adiba, Michel E., and Lindsay, B. G., "Database Snapshots," IBM Research Report RJ2772, IBM Research Laboratory, San Jose, California, March 1980. Also available in Proc. of 6th International Conf. on Very Large Databases, Montreal, October 1980.

Adiba, Michel E., "Derived Relations: a Unified Mechanism for views Snapshots and Distributed Data," IBM Research Report RJ2881, July 1980.

Adiba, Michel E., Chupin, J. C., Demolombe, R., Bihan, J. L., and Gardarin, G., "Issues in Distributed Database Management Systems: A Technical Overview," in Proceedings of the 4th International Conf. on Very Large Databases, West Berlin, September 13–15, 1978, pp. 89–110.

Agrawal, R., Carey, M. J., and Livny, M., "Models for studying concurrency control: Alternatives and implications," Proceedings of ACM-SIGMOD Conference on Management of Data, Austin, TX, SIGMOD Record, Vol. 14, No. 4, December 1985, pp. 108–119.

ALVEY, "An ALVEY Survey: Advanced information technology in the U.K.," *Future Generations Computer Systems* (FGCS), Vol. 1, No. 1, July 1984, pp. 68–78.

Allchin, J. E., and McKendry, M. S., "Synchronization and recovery of actions," Proc. Second ACM SIGACT-SOGOPS Symp. Principles of Distributed Computing, Montreal, August 1983.

Allen, F. W., Loomis, M. E. S., and Mannino, M. V., "The Integrated Dictionary—Directory System," *ACM Computing Surveys,* Vol. 14, No. 2, June 1982, pp. 245–286.

Annunziata, Robert, "Building Teleports—New York, New York," *On Communications,* Vol 2, No.5, May 1985, pp. 45–48.

ANSI X3H4 (Draft Proposed), *American National Standard Information Dictionary System: Part 1: Core Standard,* ANSI TC X3H4/85-003, American National Standards Institute, New York, 1985.

— Part 2: Entity-Level Security, ANSI TC X3H4/85-005

— Part 3: Application Program Interface, ANSI TC X3H4/85-006

— Part 4: Support of Standard Data Model, ANSI TC X3H4/85-007

ANSI (American National Standards Institute), "Data Processing Open System Interconnection—Basic Reference Model," ISO/TC97/SC16, *Computer Networks,* Vol. 5, No. 2, 1981, pp. 81–118.

ANSI/IEEE Std 729-1983, *IEEE Standard Glossary of Software Engineering Terminology,* American National Standards Institute, The Institute of Electrical & Electronic Engineering, Inc., New York, 1983.

Apers, Peter M. G., "Redundant Allocation of Relations in a Communication Network," Proc. 5th Berkeley Workshop on Distributed Data Management and Computer Networks, February 1980.

Apers, Peter M. G., "Centralized or Decentralized Data Allocation," in *Distributed Data Sharing Systems* (Van de Riet, R. P., and Litwin, W., editors), Proc. 2nd International Seminar on Distributed Data Sharing Systems, Amsterdam, North-Holland Publishing Company, June 1981.

Apers, Peter M. G., Hevner, A. R., and Yao, S. B., "Optimization Algorithms for Distributed Queries," *IEEE Transactions on Software Engineering,* Vol. SE-9, No. 1, January 1983, pp. 57–68.

Attar, Rony, Bernstein, Philip A., and Goodman, Nathan, "Site Initialization, Recovery and Back-up in a Distributed Database System," *IEEE Transactions on Software Engineering,* Vol. SE-10, No. 6, November 1984, pp. 645.

Aupperle, Eric M., "The Expert's Outlook," *IEEE Spectrum,* Vol. 22, No. 1, January 1985, pp. 36–41.

Bachman, Charles W., and Ross, Ronald G., "Toward a more complete reference model of computer-based information systems," *Computer Networks* No. 6, 1982.

Baer, Jean-Loup, "Computer Architecture," *IEEE Computer,* Vol. 17, No. 10, October 1984, pp. 77–87.

Bailey, Chris, "Hardware, Software Trends Expand Multiuser System Performance," *Mini-micro Systems,* Vol. 17, No. 8, June 15, 1984, pp. 71–76.

Balkovich, E., Lerman, S., and Parmelee, R., "Project Athena, A Joint Academic-Industry Experiment in Computers in Education," in Proceedings of the IEEE 14th Annual Conference on Frontiers in Education, Philadelphia, PA, October 3–5, pp. 466–471.

Ball, J. E., Barbacci, M. R., Fahlman, S. E., Harbison, S. P., Hibbard, P. G., Rashid, R. F., Robertson, G. G., and Steele, G. L., "The Spice Project," in *1980/1981 Computer Science Research Review,* Department of Computer Science, Carnegie-Mellon University, 1982, pp. 5–36.

Balzer, Robert, Cheatham, Thomas E., and Green, Cordell, "Software Technology in the 1990's: Using a New Paradigm," *IEEE Computer,* Vol. 16, No. 11, November 1983, pp. 39–46.

Baskett, F., Howard, J. H., and Montague, J. T., "Task Communication in DEMOS," Proc. 6th. Symposium on Operating Systems Principles, ACM SIGOPS, *Operating System Review,* Vol. 11, No. 5, November 1977, pp. 23–32.

Barnett, Richard, and Beckwith, Richard C., "Unique LAN Interconnects Diverse Computers," *Computer Design,* Vol. 22, No. 9, August 1983, pp. 77–88.

Barr, A., and Feigenbaum, E. A. (Eds.), *The Handbook of Artificial Intelligence,* Vol. 1, Stamford Heuristic Press, William Kaufman, Inc., 1981.

Bartz, Dave, "Bringing Voice Into the Long-Distance Data Network," *Infosystems,* Vol. 31, No. 11, November 1984, pp. 46–48.

Bassett, Sam, "Microprocessors: Speed Up, Price Down, and CMOS Everywhere," *Computer Design,* Vol. 22, No. 11, October 1983, pp. 177–187.

Beeby, William D., "The Heart of Integration: A Sound Database," *IEEE Spectrum,* Vol. 20, No. 5, May 1983, pp. 44–48.

Beeri, C., and Obermarck, R., "Deadlock Detection for all Resource Classes," IBM Research Report RJ3077, March 1981.

Beichter, Friedrich W., Herzog, Otthein, and Petzsch, Heiko, "SLAN-4A Software Specification and Design Language," *IEEE Transactions on Software Engineering,* Vol. SE-10, No. 2, March 1984, pp. 155–162.

Bell, Gordon C., "RISC: Back to the Future?" *Datamation,* Vol. 32, No. 11, June 1, 1986, pp. 96–108.

Bell, Trudy E., "Communication," *IEEE Spectrum,* Vol. 22, No. 1, January 1985, pp. 56–59.

Bell, Trudy E., "Communications," *IEEE Spectrum,* Vol. 23, No. 1, January 1986, pp. 49–52.

Bending, Michael J., "Hitest: A Knowledge-based Test Generation System," *IEEE Design and Testing of Computers,* Vol. 1, No. 2, May 1984, pp. 83–91.

Bennett, S. B., and Braverman, D. J., "INTELSAT VI—A Continuing Evolution," *Proceedings of the IEEE,* Vol. 72, No. 11, November 1984, pp. 1457–1468.

Bernard, Gay, "Interconnection of Local Computer Networks: Modeling and Optimization Problems," *IEEE Transactions on Software Engineering,* Vol. SE-9, No. 4, July 1983, pp. 463–470.

Bernstein, Philip A., Goodman, Nathan, Wong, Eugene, Reeve, Christopher L., and Rothnie, James B., "Query Processing in a System for Distributed Databases (SDD-1)," *ACM Transactions on Database Systems,* Vol. 6, No. 4, December 1981, pp. 602–625.

Bernstein, Philip A., and Goodman, Nathan, "Multiversion Concurrency Control-Theory and Algorithms," *ACM Transactions on Database Systems,* Vol. 8, No. 4, December 1983, pp. 465–483.

Bernstein, Philip A., and Goodman, Nathan, "Concurrency Control in Distributed Database Systems," *ACM Computing Surveys,* Vol. 13, No. 2, June 1981, pp. 185–221.

Bernstein, Philip A., and Goodman, Nathan, "An Algorithm for Concurrency Control and Recovery in Replicated Distributed Database," *ACM Transactions on Database Systems,* Vol. 9, No. 4, December 1984, pp. 596–615.

Bernstein, Philip A., Shipman, D. W., and Rothnie, J., "Concurrency Control in a System for Distributed Database (SDD-1)," *ACM Transactions on Database Systems,* Vol. 5, No. 1, March 1980, pp. 18–51.

Bernstein, Philip A., Shipman, D. W., and Wong, W. S., "Formal Aspects of Serializability in Database Concurrency Control," *IEEE Transactions on Software Engineering,* Vol. SE-5, No. 2, May 1979, pp. 203–216.

Bernstein, Philip A., and Goodman, Nathan, "A Sophisticate's Introduction to Distributed Database Concurrency Control," Proc. Eighth Conf. Very Large Databases, Mexico City, September 1982, pp. 62–76.

Bernstein, Philip A., and Goodman, Nathan, "The Failure and Recovery Problem in Replicated Databases," Proc. 2nd ACM SIGACT/SIGOPS Symposium on Principles of Distributed Computing, Montreal, Canada, August 1983.

Berry, David, "Standardizing Upper-level Network Protocols," *Computer Design,* Vol. 23, No. 2, February 1984, pp. 175–184.

Bertino, E., Haas, L. M., and Lindsay, B. G., "View Management in a Distributed Database," Proc. 9th International Conf. on very large databases, Florence, Italy, October 1983. Also available in IBM Research Report RJ3851, April 1983.

Bhandarkar, Dileep P., "Semiconductor Technology Trends and Implications," *Computer Architecture News,* Vol. 7, No. 2, August 1978.

Bhargava, Bharat, and Hua, Cecil T., "A Causal Model for Analyzing Distributed Concurrency Control Algorithms," *IEEE Transactions on Software Engineering,* Vol. SE-9, No. 4, July 1983, pp. 470–486.

Bhargava, Bharat, "Project Raid: Reliability and Integrity in Databases," Dept. Comp. Sci. Purdue University, West Lafayette, IN, January 1985.

Bhargava, Bharat, and Leslek, Lilien, "Feature Analysis of Selected Database Recovery Techniques," Dept. of Computer Science, University of Pittsburgh, 1981.

Bhargava, Bharat, and Leslek, Lilien, "Reliability in Distributed Database Systems a Survey," Tech Rep. 82-1, Dept. of Computer Science, University of Pittsburgh, 1982.

Bhargava, Bharat, "Resiliency Features of the Optimistic Concurrency Control Approach for Distributed Database Systems," Dept. of Computer Science, University of Pittsburgh, 1982.

Bhargava, Bharat, "Performance Evaluation of Reliability Control Algorithms for Distributed Database Systems," *J. Systems and Software,* Vol. 4, July 1984, pp. 239–264.

Bird, Richard, "Database Systems: All Software Versus Associative File Processors," *Computer Technology Review,* Spring-Summer 1982, pp. 55–61.

Birman, Kenneth P., Thomas, A. Joseph, Raeuchle, Thomas, and El Abbadi, Amr, "Implementing Fault Tolerant Distributed Objects," *IEEE Transactions on Software Engineering,* Vol. SE-11, No. 6, June 1985, pp. 502–508.

Birrell, A. D., Levin, R., Needham, R. M., and Schroeder, M. D., "Grapevine: An Exercise in Distributed Computing," *Communications of the ACM,* Vol. 25, No. 4, April 1982, pp. 260–273.

Birrell, A. D., and Nelson, B. J., "Implementing Remote Procedure Calls," *ACM Transactions on Computer Systems,* Vol. 2, No. 1, February 1984, pp. 39–59.

Bitton, Dina, Boral, Haran, Dewitt, David J., and Wilkinson, Kevin W., "Parallel Algorithms for the Execution of Relational Database Operations," *ACM Transactions on Database Systems,* Vol. 8, No. 3, September 1983, pp. 324–353.

Blanning, Robert W., "Conversing with Management Information Systems in Natural Language," *Communications of the ACM,* Vol. 27, No. 3, March 1984, pp. 201–207.

Boari, Maurelio, Crespi-Reghizzi, Stefano, Dapra, Alberto, Maderna, Francesco, and Natali, Antonio, "Multiple Microprocessor Programming Techniques: MML, a New Set of Tools," *IEEE Computer,* Vol. 17, No. 1, January 1984, pp. 47–61.

Boehm, B., and Barry W., *Software Engineering Economics,* Prentice-Hall, Englewood Cliffs, NJ, 1981.

Boehm, Barry W., and Standish, Thomas A., Software Technology in the 1990's: Using an Evolutionary Paradigm," *IEEE Computer,* Vol. 16, No. 1, November 1983, pp. 30–38.

Boggs, D. R., Shoch, J. F., Taft, E. A., and Metcalfe, R. M., "PUP: An Internetwork Architecture," *IEEE Transactions on Communication,* Vol. COM-28, No. 4, April 1980, pp. 612–624.

Bond, John, "CMOS Sets Pace for Chip Development," *Computer Design,* Vol. 23, No. 14, December 1984, pp. 145–153.

Booch, Grady, *Software Engineering with ADA,* Benjamin Cummings, Menlo Park, California, 1982.

Borgida, A., "Features of Languages for the Development of Information Systems at the Conceptual Level," *IEEE Software,* Vol. 2, No. 1, January 1985, pp. 63–72.

Borr, Andrea J., "Transaction Monitoring in ENCOMPASS: Reliable Distributed Transaction Processing," Tandem Technical Report TR 81.2, 1981. Also available in Proc. 7th International Conf. on Very Large Databases, Cannes, France, September 9–11, 1981, pp. 155–165.

Bourne, T. J., Gradwell, D. J. L., and Olle, T. J., "ISO/TC97/SC21/WG3, Information Resource Dictionary System working Draft Revision II," ISO, 27 June 1986.

Braden, R., Cole, R., Higginson, P., and Floyed, P., "A Distributed Approach to the Interconnection of Computer Networks," Proc. ACM SIGCOMM '83 Symp. Communication Architectures and Protocols, University of Texas at Austin, March 1983, pp. 254–259.

Branstad, Martha, and Powell, Patricia B., "Software Engineering Project Standards," *IEEE Transactions on Software Engineering,* Vol. SE-10, No. 1, January 1984, pp. 73–78.

Bray, Olin H., "Distributed Database Management: Concepts and Administration," *Journal of Telecommunication Network,* Vol. 2, No. 3, Fall 1983, pp. 237–248.

Brentano, Lewis, "Distributed CAD/CAN: Myth and Reality," *IEEE Computer Graphics and Applications,* Vol. 4, No. 8, August 1984, pp. 18–24.

Bresnahan, J. B., Barnard, D. Y., and Macleod, I. A. "WSH—A New Command Interpreter for Unix," *Software Practice and Experience,* Vol. 14, No. 12, December 1984, pp. 1197–1208.

Brill, David, Templeton, Marjorie, and Yu, Clement, "Distributed Query Processing Strategies in MERMAID, a Frontend to Data Management Systems," Proceeding of the Conference on Data Engineering, April 1984, pp. 211–218.

Brinch-Hansen, P. "Testing a Multiprogramming System," *Software Practice and Experience,* Vol. 3, No. 1, January 1973, pp. 145–150.

Brinch-Hansen, P., "The Programming Language Concurrent Pascal," *IEEE Transactions on Software Engineering,* Vol. SE-1, No. 2, February 1975, pp. 199–207.

Brodie, M. L., "The Application of Data Types To Database Semantic Integrity," *Information Systems* No. 5, 1980, pp. 287–296.

Brown, M. R., Kolling, K., and Taft, E. A., "The Alpine File System," Tech. Rep. CSL-84-4, Xerox Palo Alto Research Center, Palo Alto, CA, October 1984.

Brown, Robert L., Denning, Peter J., and Tichy, Walter F., "Advanced Operating Systems," *IEEE Computer,* Vol. 17, No. 10, October 1984, pp. 173–190.

Bryant, Susan Foster, "LAN vs Multi User Systems: Which is Best for Team Computing?" *Computer Decisions,* Vol. 16, No. 13, October 1984, pp. 112–130.

Buchanan, Bruce G., and Shortliffe, Edward H., *Rule-based Expert Systems,* Addison Wesley Publishing Company, May 1985.

Burg, Fred M., Chen, Cheng T., and Folts, Harold C., "Standards of Local Networks, Protocols, and the OSI Reference Model," *Data Communication,* Vol. 13, No. 13, November 1984, pp. 129–150.

Burr, W., "An Overview of the Proposed American National Standard for Local Distributed Data Interfaces" *Communications of the ACM,* Vol. 26, No. 10, October 1983, pp. 554–561.

Bux, W., "Performance Issues in Local Area Networks," *IBM System Journal,* Vol. 23, No. 4, December 1984, pp. 351–373.

Bux, W., Closs, F., Kuemmerle, K., Keller, H. J., and Mueller, H. R., "Architecture and Design of a Reliable Token-Ring Network," *IEEE Journal on Selected Area Communication,* SAC-1, No. 5, November 1983, pp. 756–765.

Buzbee, Bill, "Parallel Processing Makes Tough Demands," *Computer Design,* Vol. 23, No. 10, September 1984, pp. 137–140.

Caine, S., and Gordon, E., "PDL—A Tool for Software Design," Proceedings National Computer Conference, Vol. 44, ACM, 1975, pp. 271–276.

Campanella, S. J., and Harrington, J. V., "Satellite Communications Networks," *Proceeding of the IEEE,* Vol. 72, No. 11, November 1984, pp. 1506–1519.

Campbell, R., and Haberman, A., "The specification of process synchronization by path expressions," *Lecture Notes in Computer Science,* Vol. 16, 1974, pp. 89–102.

Campos, I., and Estrin, G., "Concurrent Software System Design Supported by SARA at the Age of One," Third International Conf. on Software Engineering, 1978, pp. 230–242.

Campos, I., and Estrin, G., "Concurrent Software System Design," Third International Conf. on Distributed Computer Systems, 1982, pp. 280–287.

Canada, The Federal Department of Communications "Implications of Open System Interconnection for Canada," *Canada Gazette* Part 1, Notice No. DGTP 008-85.

Canion, Rod, "Computers Reports," *PC World,* Vol. 2, No. 12, November 15, 1984, pp. 24–43.

Cardenas, Alfonso F., Alavian, Farid, and Avizienis, Algirdas, "Performance of Recovery Architectures in Parallel Associative Database Processors," *ACM Transactions on Database Systems,* Vol. 8, No. 3, September 1983, pp. 291–323.

Cardenas, Alfonso F., and Pirahesh, M. H., "Database Communication in a Heterogeneous Database Management Network," *Information System,* Vol. 5, No. 1, January 1980, pp. 55–79.

Carey, M., and Stonebreaker, M., "The performance of concurrency control algorithms for database management systems," Proceedings of the Tenth International Conference on Very Large Databases, Singapore, August 1984, pp. 107–118.

Ceri, Stefano, Navathe, Shamkant, B., and Wiederhold, Gio, "Distribution Design of Logical Database Schemas," *IEEE Transactions on Software Engineering,* Vol. SE-9, No. 4, July 1983, pp. 487–503.

Ceri, Stefano, and Pelagatti, G., "Correctness of Query Execution Strategies in Distributed Databases," *ACM Transaction on Database Systems,* Vol. 8, No. 4, December 1983, pp. 577–607.

Ceri, Stefano, Negri, M., and Pelagatti, G., "Horizontal Data Partitioning in Database Design," Proc. SIGMOD International Conference on Management of Data, June 1982.

Ceri, Stefano, and Navathe, S., "A Methodology for the Distributed Design of Databases," Proc. COMPCON, Spring 1983, San Francisco, February–March 1983.

Ceri, Stefano, and Pelagatti, Giuseppe, *Distributed Databases Principles and Systems,* McGraw-Hill Book Company, Chapter 10, 1984, pp. 277–282.

Ceri, Stefano, and Pelagatti, G., "Allocation of Operations in Distributed Database Access," *IEEE Transactions on Computers,* Vol. C-32, No. 2, February 1982, pp. 119–129.

Cerni, D., "Standards in Process: Foundations and Profiles of ISDN and OSI Studies," National Telecommunications and Information Administration, NTIA Report 84-170, December 1984.

Chamberlin, D. D., Astrahan, M. M., King, W. F., Lorie, R. A., Mehl, J. W., Price, T. G., Schkolnick, M., Selinger, P. G., Slutz, D. R., Wade, B. W., and Yost, R. A., "Support for Repetitive Transactions and Ad Hoc Queries in System R," *ACM Transactions on Database Systems,* Vol. 6, No. 1, March 1981.

Chan, Arvola, Dayal, Umeshwar, Fox, Stephen, Goodman, Nathan, Ries, Daniel R., and Skeen, Dale, "Overview of an ADA Compatible Distributed Database Manager," ACM Proceedings of SIGMOD 83, May 1983, pp. 228–237.

Chan, Arvola, Dayal, Umeshwar, Fox, Stephen, and Ries, Daniel R., "Supporting a Semantic Data Model in a Distributed Database System," Proc. 9th International Conf. on Very Large Databases, Florence, Italy, October–November 1983.

Chandy, M., Misra, J., and Haas, L., "Distributed Deadlock Detection," *ACM Transactions on Computer Systems,* Vol. 1, No. 2, May 1983.

Chang, J. M., "Simplifying Distributed Database System Design by Using a Broadcast Network," SIGMOD 84, *SIGMOD Record,* Vol. 14, No. 2, June 1984, pp. 223–233.

Chang, J. M., "A Heuristic Approach to Distributed Query Processing," Proc. of the Eighth International Conference on Very Large Database, Mexico City, VLDB Endowment, Saratoga, CA, September 1982, pp. 54–61.

Chay, Joonees K., Seltzer, Jeff, and Siddique, Naseer, "A Simple Gateway for Odd Networks," *Computer Design,* Vol. 23, No. 2, February 1984, pp. 203–212.

Chen, Bo-Shoe, and Yeh, Raymond T., "Formal Specification and Verification of Distributed Systems," *IEEE Transactions on Software Engineering*, Vol. SE-9, No. 6, November 1983, pp. 710–722.

Chen, P. P. S., "The entity-relationship model—Toward a unified view of data," *ACM Transactions on Database Systems*, Vol. 1, No. 1, March 1976, pp. 9–36.

Chen, Y. H., "PDDSCN: A Personal Distributed Database Management System on a Computer Network," *Proceedings of NCC*, 1981, Vol. 50, 1981, pp. 196–199.

Cheriton, David R., *The Thoth System: Multi-Process Structuring and Portability*, North-Holland/Elsevier, New York, 1982.

Cheriton, David R., "The V Kernel: A Software Based for Distributed Systems," *IEEE Software*, Vol. 1, No. 2, April 1984, pp. 19–42.

Cheriton, David R., and Mann, T. P., "Uniform Access to Distributed Name Interpretation in the V-System," Proc. 4th International Conf. of Distributed Computing Systems, IEEE, May 1984, pp. 290–297.

Cheriton, David R., and Roy, P. J., "Performance of the V Storage Server: A Preliminary Report," in Proceeding of the 1985 ACM 13th Annual Computer Science Conf., New Orleans, March 1985, pp. 302–308.

Cheriton, David R., and Cwaenepoel, W., "One-to-Many Interprocess Communication in the V-System," Report No. STAN-CS-84-1011, Department of Computer Science, Stanford University, Stanford, CA 94305.

Christodoulakis, Stavros, and Faloutsos, Chris, "Design Considerations for a Message File Server," *IEEE Transactions on Software Engineering*, Vol. SE-10, No. 2, March 1984, pp. 201–210.

Chu, Wesley W., and Hurley, Paul, "Optimal Query Processing for Distributed Database System," *IEEE Transactions on Computers*, Vol. C-31, No. 9, September 1982, pp. 835–850.

Chu, Wesley W., and Haverty, Patrick J., "Design Considerations for Shared Integrated Database Systems," *Journal of Telecommunication Networks*, Vol. 2, No. 3, Fall 1983.

Chu, Wesley W., "Performance of File Directory Systems for Databases in Star and Distributed Network," AFIPS Conference Proceeding, Vol. 45, NCC 1976.

Clark, D. D., Pogran, K. T., and Reed, D. P., "An Introduction to Local Area Networks," *Proc. IEEE*, Vol. 66, No. 11, November 1978, pp. 1497–1517.

Codd, E. F., "A Relational Model of Data for Large Shared Data Banks," Communication of the ACM, Vol. 13, No. 6, June 1970, pp. 377–387.

Cohen, Paul R., and Feigenbaum, Edward A. (editors), *The Handbook of Artificial Intelligence*, Vol. III, Heuristech Press, Stanford, California, and William Kaufmann, Inc., Los Altos, California, 1982.

Collie, Brian E., Kayser, Larry S., and Rybczynski, Antony M., Data Communication, *The Executive Guide to Data Communications*, Vol. 7, 1986, pp. 41–48.

Colmerauer, A., Kanoui, H., and Van Canegham, M., "Last Steps Toward an Ultimate Prolog," IJCAI-81, Vancouver, Canada, August 1981, pp. 947–948.

Colwell, Robert P., Hitchcook, I. Y. Charles, Jensen, Douglas E., Sprunt Brinkley, H. M., and Kollar, P. Charles, "Computers, Complexity, and Controversy," *IEEE Computer*, Vol. 18, No. 9, September 1985, p. 8–20.

Comer, Douglas, "The Computer Science Research Network CSNET: A History and Status Report," *Communications of the ACM*, Vol. 26, No. 10, October 1983, pp. 747–753.

Cook, Rick, "Special Report—Operating Systems," *Popular Computing*, Vol. 3, No. 10, August 1984, pp. 111–114.

Cooke, R., "Intercity Limits: Looking Ahead to All Digital Networks and No Bottlenecks," *Data Communications*, Vol. 13, No. 3, March 1984.

Cooper, Eric C., "Analysis of Distributed Commit Protocols," Proc. SIGMOD Int'l Conf. on Management of Data, June 1982, pp. 175–183.

Cooper, Eric C., "Broadband Network Design: Issues and Answers," *Computer Design,* Vol. 22, No. 3, March 1983, pp. 209–216.

Corenson, Todd L., "Integrated Software Moves Up Fast," *Mini-micro Systems,* Vol. 17, No. 7, June 1984, pp. 177–184.

Corsini, P., Frosini, G., and Lopriore, L., "Distributing and Revoking Access Authorizations on Abstract Objects: A Capability Approach," *Software Practice and Experience,* Vol. 14, No. 10, October 1984, pp. 931–943.

Cotton, I. W., "Techniques for Local Area Networks," *Computer Networks,* Vol. 4, 1980, pp. 197–208.

Craft, D. H., "Resource Management in Decentralized System," ACM Proc. 9th Symposium on Operating Systems Principles, October 1983, pp. 11–19.

Crookes, D., and Elder, J. W. G., "An Experiment in Language Design for Distributed Systems," *Software Practice and Experience,* Vol. 14, No. 10, October 1984, pp. 957–971.

Crump, Stuart Jr., "You Can Afford a Car Telephone," Future Comm Publications Inc., February 1985.

Curry, Jane, "Language-based Architecture Eases System Design—III," *Computer Design,* Vol. 23, No. 1, January 1984, pp. 127–136.

Curtis, Ronald, and Wittie, Larry, "Global Naming in Distributed Systems," *IEEE Software,* Vol. 1, No. 3, July 1984, pp. 76–80.

Daniels, D., "Query Compilation in a Distributed Database System," IBM Research Report RJ3423, March 1982.

Daniels, D., Selinger, P. G., Haas, L. M., Lindsay, B. G., Mohan, C., Walker, A., and Wilms, P., "An Introduction to Distributed Query Compilation in R*," IBM Research Report RJ3497, September 1982. Also available as proceedings 2nd International Conf. on Distributed Database, Berlin, September 1982, pp. 291–309.

Daniels, D., and Ng, P., "Query Compilation in R*," *IEEE Database Engineering,* Vol. 5, No. 3, September 1982, pp. 15–18.

Date, C. J., *An Introduction To Database Systems,* 3rd Ed., Addison-Wesley, Reading, MA, 1980.

Date, C. J., *An Introduction to Database Systems,* Volume II, Addison-Wesley, Reading, MA, 1983.

Date, C. J., *A Guide To DB2,* Addison-Wesley, Reading, MA, 1984.

Davidson, John M., "Local Network Technologies for the Office," Technical Paper Series, Paper Number 2 Ungermann-Bass Inc., 2560 Mission College Boulevard, Santa Clara, CA 95050.

Davidson, Susan B., "Optimism and Consistency in Partitioned Distributed Database Systems," *ACM Transactions on Database Systems,* Vol. 9, No. 3, September 1984, pp. 456–481.

Davidson, Susan B., and Garcia-Molina, H., "Protocols for Partitioned Distributed Database Systems," Proceedings of the Symposium on Reliability in Distributed Software and Database Systems, Pittsburgh, Pennsylvania, July 1981, pp. 145–149.

Dayal, Umeshware, "Processing Queries Over Generalization Hierarchies in a Multidatabase System," Proc. 9th International Conference on Very Large Databases, Florence, Italy, October–November 1983.

Dayal, Umeshware, Landers, T. A., and Yedwab, L., "Global Query Optimization Techniques in MULTIBASE, A System for Heterogeneous Distributed Database," Technical Report CCA-82-05, Computer Corporation of America, Cambridge, MA, 1982.

Dayal, Umeshware, Goodman, N., Landers, T. A., Olson, K., Smith, J. M., and Yedwab, L., "Local Query Optimization in MULTIBASE," Report CCA-81, Computer Corporation of America, Cambridge, MA, September 1981.

Dayal, Umeshware, and Hwang, H. Y., "View Definition and Generalization for Database Integration in MULTIBASE—A System for Heterogeneous Distributed Databases," Proceedings Sixth Berkeley Workshop on Distributed Database Management and Computer Networks, Lawrence Berkeley Laboratory, Berkeley, May 1982, pp. 203–238.

Decina, M., "CCITT Recommendations on the ISDN: A Review," *Telephony,* December 3, 1984.

Denning, Dorothy E., "Digital Signatures with RSA and Other Public-Key Cryptosystems," *Communications of the ACM,* Vol. 27, No. 4, April 1984, pp. 388–392.

Dennis, J., "Data Flow Supercomputers," *IEEE Computer,* Vol. 13, No. 11, November 1980, pp. 48–56.

Denny, Robert B., "Interne + Linkages Provide for Local Area Network Expansion," *Computer Technology Review,* Spring-Summer 1982, pp. 329–344.

Deogun, J. S., Raghavan, V. V., and Tsou, T. K. W., "Organization of Clustered Files for Consecutive Retrieval," *ACM Transactions on Database Systems,* Vol. 9, No. 4, December 1984, pp. 646–671.

Devor, Cory, Elmasri, Ramez, and Rahimi, Said, "The Design of DDTS: A Testbed for Reliable Distributed Database Management," Proc. 2nd IEEE Symposium on Reliability in Distributed Software and Database Systems, Pittsburgh, July 1982.

Devor, Cory, Elmasri, Ramez, Larson, James A., Rahimi, Said, and Richardson, James P., "Five Schema Architecture Extends DBMS to Distributed Applications," *Electronic Design,* Vol. 30, No. 5, March 18, 1982, pp. SS27–SS32.

Dijkstra, Edsger W., "The Structure of the THE-Multiprogramming System," *Communications of the ACM,* Vol. 11, No. 5, May 1968, pp. 341–346.

Dijkstra, Edsger W., "Hierarchical ordering of sequential processes," *Acta Informatica,* Vol. 1, 1971, 115–138.

Dijkstra, Edsger W., "Guarded Commands, non-determinancy and a calculus for the derivation of programs," *Communications of the ACM,* Vol. 18, No. 8, August 1975, pp. 453–457.

Dion, J., "The Cambridge File Server," *ACM Operating System Review,* Vol. 14, No. 4, October 1980, pp. 26–35.

Dorros, I., "Telephone Nets Go Digital," *IEEE Spectrum,* Vol. 20, No. 4, April 1983.

Druffel, L., Redwine, S., and Riddle, W., "The Stars Program: Overview and Retionale," *IEEE Computer,* Vol. 16, No. 11, November 1983, pp. 21–29.

Dubourdieu, D., "Survey of Current Research at Prime Computer Inc. in Distributed Database Management Systems," *Database Engineering Newsletter,* Vol. 5, No. 4, December 1982, pp. 20–22.

Dubourdieu, D., "Implementation of Distributed Transactions," Proceeding of the Sixth Berkeley Workshop on Distributed Data Management and Computer Networks, February 1982, pp. 69–90.

Duc, N., and Chew, E., "ISDN Protocol Architecture," *IEEE Communications Magazine,* March 1985.

Dunn, Lowell W., "Operating Systems Adapt to Handle Database Applications," *Computer Design,* Vol. 23, No. 4, April 1984, pp. 149–157.

Durell, William R., *Data Administration: A Practical Guide to Successful Data Management,* Chapter 2, McGraw-Hill, 1985, pp. 9–30.

Eager, Derek L., and Sevcik, Kenneth C., "Achieving Robustness In Distributed Database Systems," *ACM Transactions on Database Systems,* Vol. 8, No. 3, September 1983, pp. 354–381.

ECMA, "Second Draft, OSI Directory: Concepts and Directory Access Service," ECMA/TC23/84/159, European Computer Manufacturers Associations, Dec. 1984.

Edhart, J. L., "Understanding Local Area Networks," *Seybold Report on Word Processing,* Vol. 4, No. 6, June 1981.

Effelsberg, W., Finkelstein, S., and Schkolnick, M., "Single Database Image In a Cluster of Processors," IBM Research Report RJ4175, January 1984.

El-Ayat, Khaleo, and Agarwal, K. Rakesh, "The Intel 80386—Architecture and Implementation," *IEEE Micro,* Vol. 5, No. 6, December 1985, pp. 4–22.

Ellis, C., "Consistency and Correctness of Duplicate Database Systems," Proc. Sixth ACM Symposium on Operating Systems, Vol. 11, No. 5, Nov. 1977, pp. 67–84.

Epstein, R., and Stonebraker, M., "Analysis of Distributed Database Processing Strategies," Proc. of the Sixth International Conference on Very Large Data Bases, Montreal, IEEE, October 1–3, 1980, pp. 92–101.

Erlin, Dan, "Establishing the Micro-to-Mainframe Connection," *Computer Design,* Vol. 23, No. 2, February 1984, pp. 231–236.

Esculier, Christian, "The SIRIUS-DELTA Architecture: A Framework for Co-Operating Database Systems," *Computer Networks,* Vol. 8, No. 1, 1984, pp. 43–48.

Esculier, Christian, and Glorieus, A. M., "The SIRIUS-DELTA Distributed DBMS," (Chen, P., editor), *Entity-Relationship Approach to System Analysis and Design,* North-Holland, Amsterdam, 1980, pp. 612–624.

Esculier, Christian, "An Overview of SIRIUS, The French Pilot Project on Distributed Databases," *Journal of Telecommunication Networks,* Vol. 2, No. 3, Fall 1983.

ESPRIT, "Esprit: European Challenges U.S., and Japanese Competitors," *Future Generations Computers Systems* (FGCS), Vol. 1, No. 1, July 1984, pp. 61–68.

ESPRIT, "Esprit Projects: Year 2," Commission of the European Communities, Task Force Information Technologies and Telecommunications, Sept. 1985.

Eswaran, K. P., Gray, J. N., Lorie, R. A., and Traiger, I. L., "The notion of consistency and predicate locks in a data base system," *Communications of the ACM,* Vol. 19, No. 11, November 1976, pp. 624–633.

Fahlman, S. E., and Steele, G. L., "Tutorial on AI Programming Technology: Language and Machines," AAAI-82, Pittsburgh, PA, August 16, 1982.

Farrag, Abdel A., and Ozsu, Tamer M., "A General Concurrency Control for Database Systems," AFIPS Conference Proceeding, NCC, 1985, Chicago, Illinois, July 1985, pp. 267–274.

Ferrier, A., and Stangret, C., "Heterogeneity in the Distributed Database Management System SIRIUS-DELTA," Proc. Eighth International Conf. on Very Large Databases, Mexico City, September 1982.

Fischetti, Mark A., "MCC: An Industry Response to the Japanese Challenge" *IEEE Spectrum,* Vol. 20, No. 11, November 1983, pp. 55–56.

Fischetti, Mark A., "A Review of Progress at MCC" *IEEE Spectrum,* Vol. 23, No. 3, March 1986, pp. 76–82.

Fisher, P. S., and Slonim, Jacob, "Software Engineering: An example of Misuse," *Software Practice and Experience,* Vol. 11, No. 6, June 1981, pp. 533–539.

Fisher, P. S., Hollist, P., and Slonim, Jacob, "A Design Methodology for Distributed Databases," Proceeding of Distributed Computing, COMPCON 80, 21st IEEE Computer Society International Conf. Washington, DC, 23–25 September 1980, pp. 199–202.

Fishman, M. Y., Lai, W. K., Wilkinson, "Overview of the Jasmin Database Machine."

Folts, H. C., "Open Systems Interconnection (OSI) New International Standards Architecture and Protocols for Distributed Information Systems," *Proceeding of the IEEE,* Vol. 71, No. 12, December 1983, pp. 1331–1452.

Fontaine, James A., "Controller and Micro Team Up for Smart Ethernet Node," *Computer Design,* Vol. 23, No. 2, February 1984, pp. 215–223.

Francez, N., and Yemini, S., "A fully abstract and composable inter-task communication construct," IBM T. J. Watson Research Center, Yorktown Heights, NY, Nov. 1982.

Franta, William R., and Heath, John R., "Hyperchannel Local Network Interconnection Through Satellite Links," *IEEE Computer,* Vol. 17, No. 5, May 1984, pp. 30–39.

Freedman, David H., "New Directions in Fault-Tolerant Computers," *Infosystems,* Vol. 32, No. 7, July 1985, pp. 40–46.

Freeman, Harvey A., and Thurber, Kenneth J., "Updated Bibliography on Local Computer Networks," *Computer Architecture News,* Vol. 8, April 1980, pp. 20–28.

Fridrich, M., and Older, W., "The FELIX File Server," Proc. Eighth ACM Symposium on Operating Systems Principles, Vol. 15, No. 5, December 1981, pp. 37–44.

Fridrich, M., and Older, W., "FELIX: The Architecture of a Distributed File System," in Proceedings of the 4th International Conference on Distributed Computing Systems, San Francisco, CA, May 14–18, 1984, pp. 422–431.

Fuchi, Kazuhiro, "The Direction the FGCS Project Will Take," *New Generation Computing,* Vol. 1, No. 1, January 1983, pp. 3–9.

Fujitani, Larry, "Laser Optical Disk: The Coming Revolution in On-Line Storage," *Communications of the ACM,* Vol. 27, No. 6, June 1984, pp. 546–554.

Galler, B. I., "Concurrency Control Performance Issues," PhD Thesis, also Tech Rep CSRG-147, University of Toronto, Canada, September 1982.

Gantz, John, "White Paper on Management Life After Divestiture," *Telecommunications Products & Technology,* January 1984, pp. 4–20.

Gantz, John, "Telecommunications: An Industry in Transition," *Telecommunication Products & Technology,* Special Section in the May 1984 issue, pp. W/5-W/28.

Garcia-Molina, Hector, and Wiederhold, G., "Read-only Transactions in a Distributed Database," *ACM Transactions on Database Systems,* Vol. 7, No. 2, June 1982.

Garcia-Molina, Hector, Germand, Frank, and Kohler, Walter H., "Debugging a Distributed Computing System," *IEEE Transactions on Software Engineering,* Vol. SE-10, No. 2, March 1984, pp. 210–219.

Garcia-Molina, Hector, "Using Semantic Knowledge for Transaction Processing in a Distributed Database," *ACM Transaction on Database Systems,* Vol. 8, No. 2, June 1983, pp. 186–213.

Gehani, Narain H., "Broadcasting Sequential Processes (BSP)," *IEEE Transactions on Software Engineering,* Vol. SE-10, No. 4, July 1984, pp. 343–351.

Germano, Frank, Jr., "The Multiple-Schema Architecture DSEED: A Distributed CODASYL Prototype System," Proceedings of COMPCON F80, 1980, pp. 383–391.

Gevarter, William B., "Expert Systems: Limited but Powerful," *IEEE Spectrum,* Vol. 20, No. 8, August 1983, pp. 39–45.

Gevarter, William B., "An Overview of Artificial Intelligence and Robotics," Vol. 1—Artificial Intelligence, NBSIR 83-2799, Prepared for National Aeronautics and Space Administration, Headquarters, Washington, DC 20546, January 1984.

Giddings, Richard V., "Accommodating Uncertainty in Software Design," *Communications of the ACM,* Vol. 27, No. 5, May 1984, pp. 428–434.

Gifford, David, and Spector, Alfred, "The TWA Reservation System," *Communications of the ACM,* Vol. 27, No. 7, July 1984, pp. 649–665.

Gifford, David K., "Weighted Voting for Replicated Data," In Proceedings of the 7th Symposium on Operating System Principles, ACM SIGOPS, *Operating Systems Review,* Vol. 15, No. 5, December 1981, pp. 150–162.

Gimpleson, L.A. "An ISDN Approach to Integrated Corporate Networks", In Perry Y. (ed) Data Communications in the ISDN Era, March 1985.

Glazer, Samuel D., "Fault Tolerant Mini Needs Enhanced Operating System," *Computer Design,* Vol. 23, No. 9, August 1984, pp. 189–204.

Gligor, Virgil D., and Luckenbaugh, Gary L., "Interconnecting Heterogeneous Database Management Systems," *IEEE Computer,* Vol. 17, No. 1, January 1984, pp. 33–46.

Gligor, Virgil D., and Fong, Elizabeth, "Distributed Database Management Systems: An Architectural Perspective," *Journal of Telecommunications Networks,* Vol. 2, No. 3, November 1983, pp. 249–270.

Gligor, Virgil D., Popescu-Zeletin, Radu, "Concurrency Control Issues in Distributed Heterogenous Database Management Systems," *Distributed Data Sharing System* (F. A. Schreiber and W. Litwin, editors), 1985, pp. 43–56.

Goel, Arvind K., and Amer, Paul D., "Performance Metrics for Bus and Token-Ring Local Area Networks," *Journal of Telecommunication Networks,* Vol. 2, No. 2, Summer 1983, pp. 187–210.

Goldberg, R. P., "Architecture of Virtual Machines," Proceedings AFIPS National Computer Conference, 1972, pp. 309–318.

Goldberg, A., and Robson, D., *Smalltalk-80: The Language and its Implementation,* Addison-Wesley, Reading, MA, 1983.

Goldfine, Alan H., and Konig, Patricia A., "A Technical Overview of the Information Resource Directory System," NBSIR-85/3164 National Bureau of Standards, Gaithersburg, MD, April 1985. pp. 1–136.

Goldstein, B. C., Heller, A. R., Moss, F. H., and Wladawsky-Berger, I., "Directions in Cooperative Processing Between Workstations and Hosts," *IBM Systems Journal,* Vol. 23, No. 3, 1984, p. 236–244.

Gottlieb, Allan, and Schwartz, J. T., "Networks and Algorithms for Very-Large-Scale Parallel Computation," *IEEE Computer,* Vol. 15, No. 1, January 1982, pp. 27–36.

Government of Canada, "Trying Out the Future: Office Communication Systems in the Federal Government," Government of Canada, Department of Communications, 1984.

Graube, Maris, and Mulder, Michael, "Local Area Networks," *IEEE Computer,* Vol. 17, No. 10, October 1984, pp. 242–247.

Gray, Jim N., "A Discussion of Distributed Systems," IBM Research Report RJ2699, September 1979.

Gray, Jim N., "A Transaction Model," IBM Research Report RJ2895, August 1980.

Gray, Jim N., "The Transaction Concept: Virtues and Limitations," Proc. 7th International Conf. on Very Large Databases, October 1981, pp. 144–154.

Gray, Jim, "Practical problems in data management," Proceedings of ACM-SIGMOD 1983 Conference on Management of Data, *SIGMOD Record,* Vol. 13, No. 4, December 1983.

Gray, Jim N., Homan, P., Obermarck, Ron, and Korth, H., "A Straw Man Analysis of Probability of Waiting and Deadlock," Proc. 5th Berkeley Workshop on Distributed Data Management and Computer Networks, Lawrence Berkeley Laboratory, University of California, Berkeley, February 1981a, pp. 125–128.

Gray, Jim N., McJones, P., Blasgen, M., Lindsay, B., Lorie, R., Price, T., Putzolu, F., and Traiger, I., "The Recovery Manager of the System R Database Manager," *ACM Computing Surveys,* Vol. 13, No. 2, June 1981b, pp. 223–242.

Gray, P. D. M., "Efficient PROLOG Access to CODASYL and FDM Databases," Proceedings of ACM-SIGMOD 1985 Conference on the Management of Data, Austin, TX, *SIGMOD Record,* Vol. 14, No. 4, December 1985, pp. 437–443.

Groff, James R., "Unix Establishes a Growth Pattern," *Mini-micro Systems,* Vol. 17, No. 7, June 1984, pp. 191–197.

Gupta, Amar, Toong, Hoo-Min D., "Microprocessors—The First Twelve Years," *Proceedings of the IEEE,* Vol. 71, No. 11, November 1983, pp. 1236–1256.

Guterl, Fred, "Personal Computers," *IEEE Spectrum,* Vol. 22, No. 1, January 1985, pp. 45–49.

Guttag, J., "Abstract data types and the development of data structures," *Communications of the ACM,* Vol. 20, No. 6, June 1977, pp. 396–404.

Guttag, J., and Horning, J. J., "Formal Specifications as a Design Tool," Seventh Symposium of Principles of Programming Languages, ACM SIGPLAN/SIGACT, Jan. 1980, pp. 251–261.

Haas, Laura M., Selinger, P., Bertino, E., Daniels, D., Lindsay, B. G., Lohman, G., Masunaga, Y., Mohan, C., Ng, P., Wilms, P., and Yost, R. "R*: A Research Project on Distributed Relational DBMS," IBM Research Report RJ3653, October 1982. Also available as *Database Engineering,* Vol. 5, No. 4, December 1982, pp. 28–32.

Haas, Laura M., and Mohan, C., "A Distributed Deadlock Detection Algorithm for a Resource Based System," IBM Research Report RJ3765, January 1983.

Haatec, Eugene R., "Current Semiconductor Memories," *Computer Design,* Vol. 1, No. 4, April 1978, pp. 126–155.

Habermann, A. N., Flon, L., Cooprider, L., "Modularization and hierarchy in a family of operating systems," *Communications of the ACM,* Vol. 19, No. 5, May 1976, pp. 266–272.

Hac, Anna, "Distributed File Systems," *Operating Systems Review,* Vol. 19, No. 1, January 1985, pp. 15–18.

Haerder, Theo, and Reuter, Andreas, "Principles of Transaction-oriented Database Recovery," *ACM Computing Surveys,* Vol. 15, No. 4, December 1983, pp. 287–317.

Hammer, Michael M., "The Battle for the Desktop," *Datamation,* Vol. 30, No. 10, July 1984, pp. 68–74.

Hammer, Michael M., and Shipman, D., "Reliability Mechanisms for SDD-1: A System for Distributed Databases," *ACM Transactions on Database Systems,* Vol. 5, No. 4, December 1980, pp. 431–466.

Hattori, Takashi, and Yokoi, Toshio, "Basic Constructs of the SIM Operating System," *New Generation Computing,* Vol. 1, No. 1, 1983, pp. 81–85.

Hayes, John E. Jr., "Taking a Shot at Bypass—A BOC Copes," *On Communications,* Vol. 2, No. 6, June 1985, pp. 35–38.

Hayes-Roth, Frederick, "The Knowledge-based Expert System," *IEEE Computer,* Vol. 17, No. 9, September 1984, pp. 11–28.

Hayes-Roth, Frederick, "Knowledge-based Expert Systems," *IEEE Computer,* Vol. 17, No. 10, October 1984, pp. 263–273.

Haynes, Leonard S., Lau, Richard L., Siewiorek, Daniel P., and Mizell, David W., "A Survey of Highly Parallel Computing," *IEEE Computer,* Vol. 15, No. 1, January 1982, pp. 9–24.

Hecht, Herbert, "Computer Standards," *IEEE Computer,* Vol. 17, No. 10, October 1984, pp. 33–44.

Hedelman, Harold, "A Dataflow Approach to Procedural Modeling," *IEEE Computer Graphics and Applications,* Vol. 4, No. 1, January 1984, pp. 16–26.

Heffron, Gordon, "Teleconferencing Comes of Age," *IEEE Spectrum,* Vol. 21, No. 10, October 1984, pp. 61–67.

Held, G., Stonebraker M., and Wong, E., "Ingres: A Relational Database System," Proc. Afips 1975, NCC Vol. 44, pp. 409–416.

Hergenhan, C. B., Kretsch, D. J., and Wehr, L. A., "Unix System V Port Acceptance Criteria," Internal Document, Bell Laboratories, Murray Hill, New Jersey, 1984.

Hevner, Alan R., "Methods for Data Retrieval in Distributed Systems," Proc. Second IEEE Symposium on Reliability in Distributed Software and Database Systems, Pittsburg, July 1982.

Hillhouse, Joseph, "PABX: The Hub—Keeping Communications on Track," *Computer Decision* (Special Issue), Vol. 16, No. 15, November 1984, pp. 84–91.

Hindin, Harvey J., "Operating Systems for Minis Emphasize Unix Compatibility," *Computer Design,* Vol. 23, No. 9, August 1984, pp. 165–176.

Hindin, Harvey J., "Software Industry Strives for a Standard Solution," *Computer Design,* Vol. 23, No. 14, December 1984, pp. 217–242.

Hindin, Harvey J., "Local Networks Will Multiply Opportunities in the 1980's," *Electronics,* Vol. 55, No. 2, January 27, 1982, pp. 89–90.

Hinxman, Anthony I., "Updating a Database in an Unsafe Environment," *Communications of the ACM,* Vol. 27, No. 6, June 1984, pp. 564–566.

Hirose, Stan, and Murdock, Alan, "Network Director Service Basic Administration Service Specification," CoGnos Incorporated, Prepared for the National Library of Canada under contract OEY83-06900, September 1984.

Hoare, C. A. R., "Monitors: An Operating System Structuring Concept," *Communications of the ACM,* Vol. 15, No. 10, October 1974, pp. 549–557.

Hoare, C. A. R., "Communicating Sequential Processes," *Communications of the ACM,* Vol. 21, No. 8, August 1978, pp. 666–677.

Hogg, J., and Gamvroulas, S., "An Active Mail System," Proceeding of Annual Meeting SIGMOD '84, Vol. 14, No. 2, June 1984, pp. 215–222.

Hopper, A., "Local Area Computer Communications Network," Ph.D. Dissertation, University of Cambridge Computer Lab. Tech. Rep. 7, April 1978.

Hopper, A., "The Cambridge Ring—A Local Network," in *Advanced Techniques for Microprocessor Systems,* (Felix Hanna, editor), Peter Peregrinus, 1980.

Houghton, Raymond C., "Online Help Systems: A Conspectus," *Communications of the ACM,* Vol. 27, No. 2, February 1984, pp. 126–133.

Housel, B. C., and Scopinich, C. J., "SNA Distribution Services," *IBM Systems Journal,* Vol. 22, No. 4, December 1983, pp. 319–343.

IEEE, "ADA as a PDL," Working Group (P990) of the Technical Committee on Software Engineering of the IEEE Computer Society, 1984.

Ironman, "Requirements for High-Order Programming Languages, Ironman," United States Department of Defense, Washington, D.C., January 1977.

ISO 7498, Information Processing—Open System Interconnection Basic Reference Model, 1983.

ISO 7498, "Working Draft Addenda to ISO 7498 on Naming and Addressing," ISO TC 97/SC, 16/WG1 OSI Reference Model 180/TC/97/SC16/WG N336, June 1984.

ISO 8073, Information Processing Systems—Open Systems Interconnection—"Connection Oriented Transport Protocol Specification," 1986.

ISO DIS 8327, "Basic Connection-Oriented Session Protocol Specification," September 1984.

ISO DIS 8473, Information Processing Systems—Data Communications—"Protocol for Providing the Connectionless-Mode Network Service," March 1986.

ISO DIS 8571, "File Transfer, Access and Management (FTAM)," Parts 1 to 4, 1986.

ISO DP 8649/1, "Definition of Common Application Service Elements (CASE) Part 1: Introduction," August 1984.

ISO DIS 8649/2, "Service Definition for Common Application Service Elements (CASE) Part 2: Association Control," February 1986.

ISO DIS 8649/3, "Definition of Common Application Service Elements (CASE) Part 3: Commitment, Concurrency and Recovery," May 1985.

ISO DP 8650/1, "Specification of Protocols for Common Application Service Elements (CASE), Part 1: Introduction." August 1984.

Magnuson, Leonard H., and Szabados, Michael, "Open Net: A Network Architecture for Connecting Different Operating Systems," AFIPS Conference Proceeding, NCC 1985, Chicago, Illinois, Vol. 54, July 15–18, 1985, pp. 619–623.

Manber, Udi, and Ladner, Richard E., "Concurrency Control in a Dynamic Search Structure," *ACM Transactions on Database Systems,* Vol. 9, No. 3, September 1984, pp. 439–455.

Martin, Edith W., "The Context of Stars," *IEEE Computer,* Vol. 16, No. 11, November 1983, pp. 14–20.

Mauriello, Ralph, "Rugged Distributed Systems Adapt for Survival," *Computer Design,* Vol. 22, No. 8, July 1983, pp. 89–94.

Maxwell, Kim "High-speed Dial-up Modems," *Byte,* Vol. 9, No. 12, December 1984, pp. 179–182.

Mayberry, Walter, and Efland, Gregory, "Cache Boosts Multiprocessor Performance," *Computer Design,* Vol. 23, No. 13, November 1984, pp. 133–138.

Meister, B. W., Janson, P. A., and Svoboclova, L., "Connection-Oriented Versus Connectionless Protocols: A Performance Study," *IEEE Transactions on Computers,* Vol. C-34, No. 12, December 1985.

Menning, Walter P., Curtis, Robert E., and Burger, James, "The Information Center," *Infosystems,* Vol. 31, No. 11, November 1984, pp. 36–38.

Metcalfe, R. M., and Boggs, D. R., "Ethernet: Distributed Packet Switching for Local Computer Networks," *Communications of the ACM,* Vol. 19, 1976, pp. 395–415.

Meyer, B., "On Formalism in Specifications," *IEEE Software,* Vol. 2, No. 1, January 1985, pp. 6–26.

Misra, J., and Chandy, K., "Proofs of network processes," *IEEE Transactions on Software Engineering,* Vol. SE-7, No. 4, April 1981, pp. 417–426.

Miller, K. C., and Thompson, D. M., "Making a Case for Token Passing in Local Networks," *Data Communication,* Vol. 11, No. 3, March 1982, pp. 79–88.

Mitchell, J. G., and Dion, J., "A Comparison of Two Network Based File Servers," *Communications of the ACM,* Vol. 25, No. 4, April 1982, pp. 233–245.

Moad, Jeff, "Gambling on RISC," *Datamation,* Vol. 32, No. 11, June 1, 1986, pp. 86–92.

Mockapetris, P., "Domain Names: Concepts and Facilities," RFC 882, Network Information Center, SRI International, September 1983.

Mohan, C., Fussell, D., and Silberschatz, A., "Compatibility and Commutativity in Non-Two-Phase Locking Protocols," ACM SIGACT-SIGMOD Symposium on Principles of Database Systems, Los Angeles, March 1982.

Mohan, C., Fussell, D., and Silberschatz, A., "A Biased Non-Two-Phase Locking Protocol," Proc. International Conf. on Databases: Improving Usability and Responsiveness, Jerusalem, Israel, June 1982, published as *Improving Usability and Responsiveness* (P. Scheuermann, editor), Academic Press, NY, 1982.

Mohan, C., "An Analysis of the Design of SDD-1: A System for Distributed Databases, In Distributed Databases," *Infotech State of the Art Report,* Infotech International, England, 1979. Also Tech. Report SDBEG-11, University of Texas at Austin, April 1980.

Mohan, C., "Distributed Database Management: Some Thoughts and Analyses," Proc. ACM Annual Conf., Nashville, October 1980. Also Tech. Report TR-129, University of Texas at Austin, May 1979.

Mohan, C., *Tutorial: Recent Advances in Distributed Database Management,* IEEE Computer Society 445 Hoes Lane, Piscataway, NJ 08854, 1984.

Mohan, C., "Distributed Database Management: Some Thoughts and Analyses," Technical Report TR-129 University of Texas at Austin, May 1979.

Mohan, C., Fussell, D., and Silberschatz, A., "Compatibility and Commutativity of Lock Modes," IBM Research Report RJ3948, July 1983.

Mohan, C., Fussell, D., Kedem, Z., and Silberschatz, A., "Lock Conversions in Non-Two-Phase Protocols," IBM Research Report RJ3947, July 1983.

Mohan, C., and Lindsay, B. G., "Efficient Commit Protocols for the Tree of Processes Model of Distributed Transaction," IBM Research Report RJ3881, June 1983. Also available as Proc. 2nd ACM SIGACT-SIGOPS Symposium on Principles of Distributed Computing, Montreal, Canada, August 1983, pp. 76–88.

Mohan, C., Strong, R., and Finkelstein, S., "Method for Distributed Transaction Commit and Recovery Using Byzantine Agreement Within Cluster of Processors," IBM Research Report RJ3882, June 1983. Also available as Proc. 2nd ACM SIGACT/SIGOPS Symposium on Principles of Distributed Computing, Montreal, Canada, August 1983.

Mokhoff, Nicolas, "Networks Expand as PBSs Get Smarter," *Computer Design,* Vol. 23, No. 2, February 1984, pp. 149–168.

Mokhoff, Nicolas, "Peripherals Get Smarter as Costs Move Downward," *Computer Design,* Vol. 23, No. 14, December 1984, pp. 117–125.

Mokhoff, Nicolas, "Thirty-two Bit Micros Power Workstations," *Computer Design,* Vol. 23, No. 7, June 15, 1984, pp. 97–112.

Montgomery, W. A., "Measurements of Sharing in Multis," Proc. Sixth ACM Symposium on Operating System Principles, November 1977, pp. 85–90.

Moore, Steve, "The Mass Storage Squeeze," *Datamation,* Vol. 30, No. 10, October 1984, pp. 68–79.

Morgan, W. L., "Satellite Locations—1984," *Proceeding of the IEEE,* Vol. 72, No. 11, November 1984, pp. 1434–1444.

Moss, J., and Eliot, B., "Nested Transactions and Reliable Distributed Computing," Proc. Second IEEE Symposium on Reliability in Distributed Software and Database Systems, Pittsburgh, July 1982, pp. 33–39.

Moto-Oka, Tohru, and Stone, Harold S., "Fifth-Generation Computer Systems: A Japanese Project," *IEEE Computer,* Vol. 17, No. 3, March 1984, pp. 6–13.

Mueller, E. T., Moore, J. D., and Popek, G. J., "A Nested Transaction Mechanism for LOCUS," Proc. Ninth Symposium on Operating Systems Principles, October 11–13, 1983, ACM SIGOPS *Operating Systems Review,* Vol. 17, No. 5, October 1983, pp. 71–89.

Muntz, Rudolf, "The Distributed Database System VDN," *Database Engineering Newsletter,* Vol. 5, No. 4, December 1982, pp. 269–272.

Muntz, Rudolf, "Transaction Management in the Distributed Database System VDN," Proc. IFIP Congress 80 Tokyo/Melbourne, October 1980.

Murdock, Alan, Hirose, Stan, and Pries, Eric, "Network Directory Service Specification," CoGnos Inc. Prepared for the National Library of Canada, under contract OEY83-06900, September 1984.

Murdock, Alan, and Hirose, Stan, "Network Directory Services Management and Maintenance Guidelines," CoGnos Inc., Prepared for the National Library of Canada, under contract OEY83-02134, July 1984.

Murdock, Alan, and Hirose, Stan, "Network Directory Service Data Model," CoGnos Inc., Prepared for the National Library of Canada, under contract OEY83-06900, September 1984.

Muro, S., Ibaraki, T., Miyajima, H., and Hasegawa, T., "File Redundancy Issue in Distributed Database Systems," in Proc. 9th Int. Conf. Very Large Database, Florence, Italy, 1983, pp. 275–277.

Musa, D., "Software Engineering: The Future of a Profession," *IEEE Software,* Vol. 2, No. 1, January 1985, pp. 55–62.

McLeod, Dennis, and Heimbinger, Dennis, "A Federated Architecture for Database Systems," AFIPS Conference Proceedings, Vol. 49, National Computer Conference, 1980, pp. 283–289.

Nakanishi, T., "Correctness and Performance Evaluation of Two Phase Commit-Based Protocol for DDBS," *Computer Performance* (GB), Vol. 5, No. 1, March 1984, pp. 38–54.

Naeini, Ray, "A Few Statement Types Adapt C Language to Parallel Processing," *Electronics,* Vol. 57, No. 13, June 28, 1984, pp. 125–129.

Natarajan, N., "Communication and Synchronization Primitives for Distributed Programs," *IEEE Transactions on Software Engineering,* Vol. SE-11, No. 4, April 1985, pp. 396–416.

National Bureau of Standards, *An Overview of Expert Systems,* Gaithersburg, MD, NBS Publication NBSIR 82-2525, 1982.

Nauman, John, "ENCOMPASS: Evaluation of a Distributed Database/Transaction System," *Database Engineering Newsletter,* Vol. 5, No. 4, December 1982, pp. 37–41.

Navathe, Shamkant, B., Ceri, S., Wiederhold, G., and Dou, J., "Vertical Partitioning for Physical and Distribution Design of Databases," Technical Report STAN-CS-82-957, Stanford University, January 1983.

Neches, Philip M., "Hardware Support for Advanced Data Management Systems," *IEEE Computer,* Vol. 17, No. 11, November 1984, pp. 29–41.

Needham, R. M., and Herbert, A. J., *The Cambridge Distributed Computing System,* Addison Wesley, Reading, MA, 1982.

Neigh, James L., "Digital Multiplexed Interface: Architecture and Specifications," AFIPS Conference Proceeding, NCC 1985, Chicago, Illinois, Vol. 54, July 1985, pp. 633–638.

Neuhold, E. J., and Walter, B., "The DDBS POREL: Current Research Issues and Activities," *Database Engineering Newsletter,* Vol. 5, No. 4, December 1982, pp. 278–282.

Neuhold, E. J., and Walter, B., "An Overview of the Architecture of the Distributed Data Base System POREL," Proc. Second Intenational Symposium on Distributed Data Bases, Berlin, September 1982, pp. 247–290.

Ng, Pui, "Distributed Compilation and Recompilation of Database Queries," IBM Research Report RJ3375, January 1982.

Nguyen, G. T., "Decentralized Dynamic Query Decomposition for Distributed Database Systems," Proc. ACM Pacific '80: Distributed Processing—New Directions for a New Decade, San Francisco, November 1980.

Ni, M. Lionel, Xu Chong-Wei, and Gendreau, Thomas B., "A distributed Drafting Algorithm for Load Balancing," *IEEE Transaction on Software Engineering,* Vol. SE-11, No. 10, Oct. 1985, pp. 1153–1161.

Nilsson, N. J., "Artificial Intelligence: Engineering, Science or Slogan," *AI Magazine,* Vol. 3, No. 1, Winter 81/82, pp. 2–9.

Nissen, J. C. D., Wallis, P. J., and Wichmann, B. A., "ADA Europe Guidelines for the Portability of ADA programs," *ACM ADA Letters,* Vol. 2, No. 3, March–April 1982, pp. 44–61, also NCL Rep. DNACS 52/81, National Physical Laboratory, Teddington, England, 1981.

Nissen, J. C. D., Wallis, P. J. L., Wichmann, B. A., et al., *ADA-Europe Guidelines for the Selection and Specification of ADA Compilers,* NPL Rep. DITC 10/82, National Physical Laboratory, Teddington, England, 1982, and *ACM ADA Letters,* Vol. 3, No. 1, July–August 1983, pp. 37–50.

Noe, J., and Nutt, G., "Macro E-nets for representation of parallel systems," *IEEE Transactions on Computers,* Vol. C22, August 1973, pp. 718–727.

Noe, J., "Hierarchical modelling with pro-nets," *Proceedings of the National Electronics Conference,* Vol. 32, Oct. 1978, 155–160.

Norman, Alan, and Anderton, Mark, "EMPACT: A Distributed Database Application,"

AFIPS Conference Proceedings, NCC, Anaheim, California, Vol. 52, May 1983, pp. 203–217.

Norman, D., "Design principles for human-computer interaction," Proceedings Computer-Human Interface 83, ACM, 1983, pp. 1–10.

Novak, Robert E., "Segmented Architectures Hamper High-level Languages," *Mini-micro System,* Vol. 14, No. 12, December 1981, pp. 139–148.

Nutt, G., "Evaluation Nets for Computer System Performance Analysis," *Proceedings 1972 Fall Joint Computer Conference,* Vol. 41, 1972, pp. 279–286.

Oakley, B. W., "The ALVEY Programme: Progress Report 1985," Private note.

Obermarck, Ron, "Global Deadlock Detection Algorithm," IBM Research Report RJ2845, June 1980.

Obermarck, Ron, "Distributed Deadlock Detection Algorithm," *ACM Transactions on Database Systems,* Vol. 7, No. 2, June 1982, pp. 187–208.

Oh, Se-Young, "A Walsh-Hadamard Based Distributed Storage Device for the Associative Search of Information," *IEEE Transactions on Pattern Analysis and Machine Intelligence,* Vol. PAMI-6, No. 5, September 1984, pp. 617–623.

Oppen, Derek C., and Yogen, Dalal K., "The Clearinghouse: A Decentralized Agent for Locating Named Objects in a Distributed Environment," *ACM Transactions on Office Information Systems,* Vol. 1, No. 3, July 1983, pp. 230–253.

Owicki, S., and Gries, D., "Verifying Properties of Parallel Programs: An Axiomatic Approach," *Communications of the ACM,* Vol. 19, No. 1, May 1976, pp. 279–285.

Owicki, S., and Lamport, L., "Proving liveness properties of concurrent programs," *ACM Transactions on Programming Languages and Systems,* Vol. 4, July 1982, pp. 455–495.

Panzieri, F., and Randell, B., "Interfacing UNIX to Data Communications Networks," *IEEE Transactions on Software Engineering,* Vol. SE-11, No. 10, October 1985, 1016–1031.

Papadimitriou, C. H., "The Serializability of Concurrent Database Updates," *Journal of the ACM,* Vol. 26, No. 4, October 1979, pp. 631–653.

Parr, F. N., and Strom, R. E., "NIL: A High-level Language for Distributed Systems Programming," *IBM Systems Journal,* Vol. 22, Nos. 1/2, 1983, pp. 111–127.

Pasik, Alexander, and Schor, Marshall, "Table Driven Rules in Expert Systems," *ACM SIGART Newsletter,* No. 87, January 1984, pp. 31–32.

Patterson, D., and Sequin, C., "A VLSI RISC," *IEEE Computer,* Vol. 15, No. 9, September 1982, pp. 8–20.

Patterson, David A., "Reduced Instruction Set Computers," *Communications of the ACM,* Vol. 28, No. 1, January 1985, pp. 8–21.

Persky, G., Aquila, D., Slonim, Jacob, Kolpowska, M., and Macrae, L. J., "A GEAC Local Area Network for the Bobst Library," *Library Hi Tech.,* Vol. 2, No. 2, 1984, pp. 37–45.

Peterson, J., "Petri Nets," *ACM Computing Surveys,* Vol. 9, No. 3, Sept. 1977, pp. 223–252.

Peterson, J., *Petri Net Theory and the Modelling of Systems,* Prentice-Hall, Englewood Cliffs, NJ, 1981.

Phillips, David, "The Z80000 Microprocessor," *IEEE Micro,* Vol. 5, No. 6, December 1985, pp. 23–36.

Pnueli, A., "The temporal logic of programs," Proc. of the 18th IEEE Symposium on Foundations of Computer Science, Nov. 1977, pp. 46–57.

Pnueli, A., and DeRoever, W., "Rendezvous with Ada—A proof theoretic view," Proc. AdaTEC Conference on Ada, Arlington, VA, 1982, pp. 129–137.

Popek, Gerald T., and Kline, C. S., "Encryption and Secure Computer Networks," *Computing Surveys,* Vol. 11, No. 4, December 1979, pp. 331–356.

Popek, Gerald T., Walker, R. B., Chow, J., Edwards, D., Kline, C., Rudisin, G., and Thiel, G., "LOCUS: A Network Transparent, High Reliability Distributed System," Proc. Eighth ACM Symposium on Operating Systems Principles, December 14–16, 1981, *ACM SIGOPS Operating Systems Review,* Vol. 15, No. 5, December 1981, pp. 169–177.

Popek, Gerald T., "Notes on Distributed System of Microprocessors," University of Dublin, June 1981, pp. 287.

Pouzin, L., and Zimmerman, H., "A Tutorial on Protocols," Proc. IEEE 66, Vol. 11, November 1978, pp. 1346–1370.

Powell, M. L., "The DEMOS File System," Proc. Sixth ACM Symposium on Operating Systems Principles, *ACM SIGOPS Operating System Review,* Vol. 11, No. 5, November 1977, pp. 33–42.

Powell, M. L., and Miller, B. P., "Process Migration in DEMOS/MP," Proc. Ninth Symposium on Operating Systems Principles, *ACM SIGOPS Operating System Review,* Vol. 17, No. 5, October 1983, pp. 110–119.

Poole, F., "Microcomputers and Distributed Database Management Systems," *Data Processing* (GB), Vol. 26, No. 6, July–August 1984, pp. 20–21.

Qadah, Ghassan Z., "Database Machines: A Survey," AFIPS Conference Proceeding, NCC 1985, Chicago, Illinois, Vol. 54, July 1985, pp. 211–224.

Quarterman, John S., Silberschatz, Abraham, and Peterson, James L., "4.2 BSO and 4.3 BSO as Examples of the Unix Systems," *ACM Computing Surveys,* Vol. 17, No. 4, December 1985, pp. 379–418.

Radicati, Siranush, "Managing Transient Internetwork Links in the Xerox Internet," *ACM Transactions on Office Information Systems,* Vol. 2, No. 3, July 1984, pp. 213–225.

Rahimi, Said, Spinrad, M. D., and Larson, James A., "A Structural View of Honeywell's Distributed Database Testbed System: DDTS," *Database Engineering,* Vol. 5, No. 2, December 1982, pp. 283–287.

Ramamoorthy, C. V., Prakash, Atul, Tsai, Wei-Tek, and Usuda, Yutaka, "Software Engineering Problems and Perspectives," *IEEE Computer,* Vol. 17, No. 10, October 1984, pp. 191–211.

Ramamritham, Krithivasan, and Keller, Robert M., "Specification of Synchronizing Processes," *IEEE Transactions on Software Engineering,* Vol. SE-9, No. 6, November 1983, pp. 722–733.

Ramamritham, Krithivasan, and Stankivic, John, "Dynamic Task Scheduling in Hard Real Time Distributed Systems," *IEEE Software,* Vol. 1, No. 3, July 1984, pp. 65–75.

Rappaport, David M., "Local Area Networks: Links to Communicating," *The Office,* Vol. 97, No. 2, February 1983, pp. 103–124.

Rashid, R. F., "An Inter Process Communication Facility for UNIX," Tech. Rept. CMU-CS-80-124, Dept. of Computer Science, Carnegie-Mellon University, March 1980.

Rauch-Hindin, W., "Distributed Databases," *System & Software,* Vol. 2, No. 3, September 1983.

Reed, D. P., and Svobodova, L., "SWALLOW: A Distributed Data Storage System for a Local Network," in *Local Networks for Computer Communications,* (A. West and P. Jones, editors), North-Holland Pub., Amsterdam, 1981, pp. 355–373.

Reisner, P., "Formal Grammar and Human Factors Design for Interactive Graphics Systems," *IEEE Transactions on Software Engineering,* Vol. SE-7, No. 2, February 1981, pp. 229–240.

Rich, Elaine, "Natural-Language Interfaces," *IEEE Computer,* Vol. 17, No. 9, September 1984, pp. 39–50.

Rich, Elaine, "The Gradual Expansion of Artificial Intelligence," *IEEE Computer,* Vol. 17, No. 5, May 1984, pp. 4–14.

Riddle, W., "An Assessment of DREAM," *Software Engineering Environments* (Hunke H., ed.) North-Holland Pub. Co., Amsterdam, 1980.

Riddle, W., Wileden, J., Saylor, J., Segal, A., and Stavely, A., "Behavior modelling during software design," *IEEE Transactions on Software Engineering,* Vol. SE-4, No. 4, April 1978, pp. 283–292.

Rietmann, Kearney, "Networks Reports," *PC World,* November 15, 1984, pp. 162–171.

Rivest, Ronald L., and Shamir, Adi, "How to Expose an Eavesdropper," *Communications of the ACM,* Vol. 27, No. 4, April 1984, pp. 393–395.

Rosenberg, Robert, and Feldt, Terry E., "Local Nets Arrive in Force," *Electronics Week,* Vol. 57, No. 18, August 6, 1984, pp. 67–73.

Rosenkrantz, D. J., Stearns, R. E., and Lewis, P. M., "System level concurrency control for distributed systems," *ACM Transactions on Database Systems,* Vol. 3, No. 2, June 1978.

Rothnie, James B., Bernstein, P. A., Fox, S., Goodman, N., Hammer, M., Landers, T. A., Reeve, C., Shipman, D. W., and Wong, E., "Introduction to a System for Distributed Databases (SDD-1)," *ACM Transactions on Database Systems,* Vol. 5, No. 1, March 1980, pp. 1–17.

Roubine, O., and Robinson, L., "SPECIAL Reference Manual," Stanford Research Institute, Menlo Park, CA, 1977.

Rushby, John, and Randell, Brian, "A Distributed Secure System," *IEEE Computer,* Vol. 16, No. 7, July 1983, pp. 55–67.

Sacco, G. M., "Distributed Query Evaluation in Local Area Network," *IEEE Data Engineering,* 1984, pp. 510–516.

Sakamoto, R., "CCITT Standards Activity: the Integrated Services Digital Network (ISDN)," McLean, VA, Mitre Technical Report, MTR-82WOD69, September 1982.

Saltzer, J. H., "On the Naming and Binding of Network Destinations," Proceedings IFIPS/TC6 International Symposium on Lock Computer Networks, Florence, Italy, April 19–21, 1982, pp. 311–317.

Sandberg, R., Goldberg, D., Kleinman, S., Walsh, D., and Lyon, B., "Design and Implementation of the Sun Network File System," USENIX Summer 1985 Conference, pp. 119–130.

Sapronov, Walt, "Technical and Regulatory Issues are Challenging ISDN's Progress," *Data Communication,* Vol. 14, No. 12, November 1985, pp. 265–274.

Sastry, A., "Performance Objectives for ISDNS," *IEEE Communications Magazine,* January 1984.

Satyanarayanan, M., "The ITC Project: an Experiment in Large Scale Distributed Personal Computing," in Proceedings of the Networks 84 Conference, Madras, India, October 1984.

Saxton, W. A., and Edwards, Morris, "Decision Model for Distributed Processing," *Infosystems,* Vol. 25, No. 9, September 1978, pp. 88–91.

Schindler, Max, "Mini Operating Systems Adapt To Multi-user Demands," *Electronic Design,* Vol. 30, No. 10, May 13, 1982, pp. SS26–SS31.

Schoeffler, James D., "Distributed Computer Systems for Industrial Process Control," *IEEE Computer,* Vol. 17, No. 2, February 1984, pp. 11–18.

Schreiber, F. A., and Martella, G., "A Data Dictionary for Distributed Database" *Conf. on Distributed Database,* (C. Delobel and W. Litwin, editors), North-Holland, 1980.

Schroeder, M. D., Clark, D. D., and Saltzer, J. H., "The Multics Kernel Design Project," Proc. Sixth ACM Symposium on Operating System Principles, November 1977, pp. 43–56.

Scown, Susan J., *The Artificial Intelligence Experience: An Introduction,* Digital Equipment Corporation, May 1985.

Seaman, John, "LANs Ready to Deliver," *Computer Decisions,* Vol. 15, No. 6, June 1983, pp. 134–150.

Seaman, John, "Artificial Intelligence: Out of the Lab, Into the Workplace," *Computer Decisions,* Vol. 16, No. 10, August 1984, pp. 98–113.

Selinger, Patricia G., and Adiba, M., "Access Path Selection in Distributed Database Management Systems," IBM Research Report J2883, 1980. Also available as Proceedings of the International Conf. on Databases, Deen and Hammersly (editors), University of Aberdeen, July 1980, pp. 204–215.

Serlin, Omri, "Fault Tolerant Systems in Commercial Application," *IEEE Computer,* Vol. 17, No. 8, August 1984, pp. 19–30.

Serlin, Omri, "MIPS Dhrystones, and other tales," *Datamation,* Vol. 32, No. 11, June 1, 1986, pp. 112–118.

Sevcik, K. C., "Comparison of Concurrency Control Methods Using Analytic Models," *Proc. Information Processing 83* (Mason, R. E. A., editor), North-Holland, Paris, September 1983.

Shafer, Timothy C., "Digital Multiplexed Interface: A Host Side Implementation," AFIPS Conference Proceeding, NCC 1985, Chicago, Illinois, Vol. 54, July 1985, pp. 639–644.

Shapiro, Ehud, and Takeuchi, Akikazu, "Object Oriented Programing in Concurrent Prolog," *New Generation Computing,* Vol. 1, No. 1, January 1983, pp. 25–48.

Shapiro, Ehud Y., "The Fifth Generation Project—A Trip Report," *Communications of the ACM,* Vol. 26, No. 9, September 1983, pp. 637–641.

Shipman, D., "The functional data model and the data language DAPLEX," *ACM Transactions on Database Systems,* Vol. 6, No. 1, March 1981, pp. 140–173.

Shneiderman, B., "Human factors experiments in designing interactive systems," *IEEE Computer,* Vol. 12, No. 12, December 1979, pp. 9–19.

Shoch, John F., "Inter-Network Naming, Addressing and Routing," *Tutorial: Distributed Processor Communication Architecture* (Kenneth J. Thurber, editor), IEEE First Conf. on Distributed Computing Systems, Huntsville, Alabama, October 1979, pp. 280–287. Also available Proceedings 17th IEEE Computer Society International Conf. COMPCON September 1978, pp. 72–79.

Sidney, Fernbach, "Supercomputers, Past, Present, Prospects," FGCS, Vol. 1, No. 1, 1984, pp. 23–30.

Skeen, Dale, "Crash Recovery in a Distributed Database System," PhD Thesis, Memorandum No. UCB/ERL M82/45 University of California at Berkeley, May 1982. Also available in the Proceeding of the 6th Berkeley Workshop on Distributed Data Management and Computer Networks, February 1982, pp. 69–80.

Skeen, Dale, and Stonebraker, M. "A Formal Model of Crash Recovery in a Distributed System," *IEEE Transactions on Software Engineering,* Vol. SE-9, No. 3, May 1983.

Shemer, Jack, and Neches, Philip, "The Genesis of a Database Computer," *IEEE Computer,* Vol. 17, No. 11, November 1984, pp. 42–56.

Siewidrek, Daniel P., "Architecture of Fault-Tolerant Computers, *IEEE Computer,* Vol. 17, No. 8, August 1984, pp. 9–18.

Silberschatz, Abraham, "Cell: A Distributed Computing Modularization Concept," *IEEE Transactions on Software Engineering,* Vol. SE-10, No. 2, March 1984, pp. 178–185.

Slonim, Jacob, Macrae, L. J., Mennie, W. E., and Diamond, Norman, "NDX-100: An Electronic Filing Machine for the Office of the Future," *IEEE Computer,* Vol. 14, No. 5, May 1981, pp. 24–36.

Slonim, Jacob, Macrae, L. J., McBride, R. A., Maryanski, F. J., Unger, E. A., and Fisher, P. S., "A Throughput Model: Sequential vs Concurrent Processing," *Information Systems,* Vol. 7, No. 1, January 1982, pp. 65–83.

Slonim, Jacob, Schmidt, D., and Fisher, P., "Considerations for Determining the Degrees of Centralization or Decentralization in the Computing Environment," *Information and Management,* Vol. 2, No. 1, 1979, pp. 15–29.

Smith, Reid G., Lafue, Gilles M. E., Schoen, Eric, and Vestel, Stanley C., "Declarative Task Description as a User Interface Structuring Mechanism," *IEEE Computer,* Vol. 17, No. 9, September 1984, pp. 29–38.

Smith, John Miles, et al., "MULTIBASE-Integrating Heterogeneous Distributed Database Systems," Proceedings of AFIPS 1981, National Computer Conference, Vol. 50, 1981, pp. 487–499.

Snead, Bob, Ho, Frank, and Engram, Bob, "Operating System Features Real Time and Fault Tolerance," *Computer Design,* Vol. 23, No. 9, August 1984, pp. 177–188.

Snyders, Jan, "Let's Talk DBMS," *Infosystems,* Vol. 31, No. 12, December 1984, pp. 36–44.

Snyder, Lawrence, "Parallel Programming and The Poker Programming Environment," *IEEE Computer,* Vol. 17, No. 7, July 1984, pp. 27–37.

Sollins, K., "Distributed Name Management," Ph.D Th. Massachusetts Institute of Technology, 1985.

Solomon, M., Landweber, L., and Neuhengen, D., "The CSNET Name Server," *Computer Networks,* Vol. 6, No. 3, July 1982, pp. 161–172.

Spaccapietra, A., "Distributed DBMS Architectures," Proc. 2nd International Seminar on Distributed Data Sharing Systems, Amsterdam, June 1981.

Spaccapietra, A., Demo, B., Dileva, A., and Parent, C., "SCOOP: A System for Cooperation Between Existing Heterogeneous Distributed Data Bases and Programs," *Database Engineering Newsletter,* Vol. 5, No. 4, December 1982, pp. 52–57.

Stallings, W., "Local Networks," *ACM Computing Surveys,* Vol. 16, No. 1, March 1984.

Stallings, W., "The Evolution of Integrated Services Digital Networks," in *Oxford Surveys in Information Technology,* Vol. II, Oxford University Press 1985.

Staniszkis, W., Kaminski, W., Kowalewski, M., Krajewski, K., Mezyk, S., and Turco, G., "Network Data Management System Architecture," Position Papers of the 3rd Seminar on Distributed Data Sharing Systems, Parma, Italy, March 1984.

Stankovic, J., "Debugging commands for a distributed processing system," Proc. COMPCON Fall 1980, Vol. 25, pp. 701–705.

Stavely, Allan M., "Modeling and Projection in Software Development," *The Journal of Systems and Software,* Vol. 3, No. 2, June 1983, pp. 137–146.

Stonebraker, Michael, Woodfill, J., Ranstrom, J., Murphy, M., Kalash, J., Carey, M., and Arnold, K., "Performance Analysis of Distributed Data Base Systems," *Database Engineering Newsletter,* Vol. 5, No. 4, December 1982, pp. 58–65.

Stonebraker, Michael, "Concurrency Control and Consistency of Multiple Copies of Data in Distributed INGRES," *IEEE Transactions on Software Engineering,* Vol. SE-5, No. 3, May 1979, pp. 188–194.

Stonebraker, Michael, "Operating System Support for Database Management," *Communications of the ACM,* Vol. 24, No. 7, July 1981, pp. 412–418.

Storey, Thomas F., "DMI: A PBX Perspective," AFIPS Conference Proceeding, NCC 1985, Vol. 54, Chicago, Illinois, July 1985, pp. 625–632.

Stoy, J., *Denotational Semantics: The Scott-Strachey Approach to Programming Language Theory,* The MIT Press, Cambridge, MA, 1977.

Stuck, B. W., and Arthurs, E., "A Computer and Communications Network Performance Analysis Primer," Prentice-Hall, Englewood Cliffs, NJ, 1985.

Sturgies, H. E., Mitchell, J. G., and Israel, J. E., "Issues in the Design and Use of a Distributed File System," *ACM SIGOPS Operating System Review,* Vol. 14, No. 3, July 1980, pp. 55–69.

Su, Stanley Y. W., "A Microcomputer Network System for Distributed Relational Databases: Design, Implementation and Analysis," *Journal of Telecommunication Networks,* Vol. 2, No. 3, Fall 1983.

Suenderman, Jan A. F., "Software/Downloading Entire Modules can Reduce Server Demands," *Data Communication,* Vol. 13, No. 13, November 1984, pp. 155–164.

Summers, R. C., "An Overview of Computer Security," *IBM Systems Journal,* Vol. 23, No. 4, December 1984, pp. 309–325.

Sundstrom, R. J., Staton, J. B., Schultz, G. D., Hess, M. L., Deaton, G. A., Cole, L. J., and Amy, R. M., "SNA Directions—A 1985 Perspective," AFIPS Conference Proceeding, NCC 1985, Vol. 54, Chicago, Illinois, July 1985, pp. 589–603.

Suydam, Bill, "Rash of New Compilers Brings Multiprocessors into the ADA Fold," *Computer Design,* Vol. 25, No. 1, January 1, 1986, pp. 41–45.

Svobodova, Liba, "Resilient Distributed Computing," *IEEE Transactions on Software Engineering,* Vol. SE-10, No. 3, May 1984, pp. 257–268.

Svobodova, Liba, "File Servers for Network Based Distributed Systems," *ACM Computing Surveys,* Vol. 16, No. 4, December 1984, pp. 353–398.

Sze, D. T. W., "A Metropolitan Area Network," *IEEE Journal on Selected Areas in Communications,* Vol. SAC-3, No. 6, November 1985.

Szewerenko, Leland, Dietz, William B., and Word, Frank E., "Nebula: A New Architecture and Its Relationship to Computer Hardware," *IEEE Computer,* Vol. 14, No. 2, February 1981, pp. 35–41.

Tabak, Daniel, "Dynamic Architecture and LSI Modular Computer Systems," *IEEE Micro,* Vol. 4, No. 2, April 1984, pp. 48–66.

Takizawa, Mokoto, "Heterogeneous Distributed Database System: JDDBS," *Database Engineering Newsletter,* Vol. 6, No. 1, March 1983, pp. 58–62.

Tanenbaum, Andrew S., *Computer Networks,* Prentice-Hall, 1981, Chap. 7.

Tanenbaum, Andrew S., and Mullender, S. J., "The Design of a Capability-Based Distributed Operating System," *Computer Journal,* 1986.

Tanenbaum, Andrew S., "Network Protocols," *ACM Computing Surveys,* Vol. 13, No. 4, December 1981, pp. 453–489.

Tasaka, S., "Multiple-Access Protocols for Satellite Packet Communication Networks: A Performance Comparison," *Proceeding of the IEEE,* Vol. 72, No. 11, November 1984, pp. 1573–1582.

Tay, Y., and Suri, R., "Choice and performance in locking for databases," Proceedings of the Tenth International Conference on Very Large Databases, Singapore, August 1984.

Templeton, M., Brill, D., Hwang, A., Kameny, I., and Lund, E., "An Overview of the MERMAID System—A Frontend to Heterogeneous Databases," The Proceedings of EASCON 83, 1983, pp. 387–402.

Terry, Douglas B., "An Analysis of Naming Conventions for Distributed Computer Systems," Proc. SIGCOMM 84 Symposium on Communication Architectures and Protocols, ACM, June 1984, pp. 218–232.

Terry, Douglas B., and Andler, Sten, "The Cosie Communication Subsystem: Support for Distributed Office Applications," *ACM Transactions on Office Information Systems,* Vol. 2, No. 2, April 1984, pp. 79–95.

Thomas, R. H., "A Majority Consensus Approach to Concurrency Control for Multiple Copy Databases," *ACM Transactions on Database Systems,* Vol. 4, No. 2, June 1979, pp. 180–209.

Thurber, Kenneth J., and Freeman, Harvey A., "Architecture Considerations for Local Computer Networks," Proceedings of the First International Conference on Distributed Computing System, October 1979, pp. 131–142.

Toense, Robert E., "Performance Analysis of NBSNET," *Journal of Telecommunication Networks,* Vol. 2, No. 2, Summer 1983, pp. 177–186.

Treu, S., "Position Papers: Topical Profiles for SIGOA," *ACM SIGOA Newsletter,* Vol. 4, No. 1, Summer 1983, p. 9–12.

Thomas, R. H., "A Majority Consensus Approach to Concurrency Control for Multiple Copy Database," *ACM Transactions on Database Systems,* Vol. 4, No. 2, 1979, pp. 180–209.

Traiger, Irving L., Gray, J. N., Galtieri, C. A., and Lindsay, B., "Transactions and Consistency in Distributed Database Systems," *ACM Transactions on Database Systems,* Vol. 7, No. 3, September 1982, pp. 323–342.

Traiger, Irving L., "Trends in Systems Aspects of Database Management," IBM Research Report RJ3845, April 1983. Also available as Proc. 2nd International Conference on Database, (S. M. Deen and P. Hammersley, editors), Cambridge, England, September 1983.

Traiger, Irving L., "Virtual Memory Management for Database Systems," *ACM Operating Systems Review,* Vol. 16, No. 4, 1982, pp. 26–48.

Tsichritzis, Dionysios C., "Message Addressing Schemes," *ACM Transactions on Office Information Systems,* Vol. 2, No. 1, January 1984, pp. 58–77.

Tsichritzis, Dionysios C., and Klug, A., "The ANSI/X3/SPARC DBMS Framework Report of the Study Group on Database Management Systems," AFIPS Press, 210 Summit Ave., Montvale, N.J. 07645, 1977.

Tsichritzis, D. C., and Lochovsky, F. H., *Data Models,* Prentice-Hall, Englewood Cliffs, NJ, 1982.

Tsubaki, M., and Hotaka, R., "Distributed Multi-Database Environment with a Supervisory Data Dictionary Database," *Entity-Relationship Approach to System Analysis and Design* (Chen, P., editor), North-Holland, Amsterdam 1980, pp. 625–646.

Tsujino, Y., Ando, M., Araki, T., and Tokura, B., "Concurrent C: A Programming Language for Distributed Multiprocessor Systems," *Software Practice and Experience,* Vol. 14, No. 11, November 1984, pp. 1061–1078.

Tucker, Allen B., "A Perspective on Machine Translation: Theory and Practice," *Communications of the ACM,* Vol. 27, No. 4, April 1984, pp. 322–329.

Unger, E. A., McBride, R. A., Slonim, Jacob, and Maryanski, F. J., "Design for Integration of a DBMS into a Network Environment," Proc. of the Sixth Data Communications Symposium, Pacific Grove, CA, November 27–29, 1979, pp. 26–34.

Unger, E. A., Fisher, P. S., and Slonim, Jacob, "Evolving to Distributed Database Environments," *Computer Communications,* Vol. 5, No. 1, January 1982, pp. 17–22.

UNIX Programmer's Manual—4.2, Berkeley Software Distribution Virtual VAX-11 Version, Division Computer Science, Dept. Elec. Eng. Comp. Sci., University California, Berkeley, August 1983.

Valdorf, G., "Dedicated Distributed and Portable Operating Systems: A Structuring Concept," *Software Practice and Experience,* Vol. 14, No. 11, November 1984, pp. 1079–1093.

Vigder, Mark, and MacKinnon, D. A., "Information Search and Transfer Reference Model, Version 1.0," Computer Gateways Inc., Prepared for the National Library of Canada under contract OEU84-03067, December 1984.

Voss, K., "Using predicate/transition nets to model and analyze distributed database systems," *IEEE Transactions on Software Engineering,* Vol. SE-6, No. 11, November 1980, pp. 539–544.

Vroly, K., and John, R., "Fitting the Pieces into a Super-Minicomputer," *Mini-micro Systems,* Vol. 17, No. 13, November 1984, pp. 151–159.

Voydock, Victor L., and Kent, Stephen T., "Security Mechanisms in High Level Network Protocols," *ACM Computing Surveys,* Vol. 15, No. 2, June 1983, pp. 135–171.

Vuong, Son T., and Cowan, Donald D., "Protocol Validation via Resynthesis and Decomposition: The X.75 Packet Level as an Example," *Journal of Telecommunication Networks,* Vol. 2, No. 2, Summer 1983, pp. 153–176.

Wah, Benjamin W., "File Placement on Distributed Computer Systems," *IEEE Computer,* Vol. 17, No. 1, January 1984, pp. 23–32.

Walker, R. B., Popek, Gerald T., English, R., Kline, C., and Thiel, G., "The LOCUS Distributed Operating System," Proc. Ninth Symposium on Operating Systems Principles, October 11–13, 1983, pp. 49–70.

Walker, Henry M., "Administering a Distributed Database Management System," *ACM SIGMOD Record,* Vol. 12, No. 3, April 1982, pp. 86–99.

Wallace, John J., and Barnes, Walter W., "Designing for Ultra-high Availability: The Unix RTR Operating System," *IEEE Computer,* Vol. 17, No. 8, August 1984, pp. 31–39.

Wallich, Paul, "Minis and Mainframes," *IEEE Spectrum,* Vol. 22, No. 1, January 1985, pp. 42–44.

Wallich, Paul, "Software," *IEEE Spectrum,* Vol. 22, No. 1, January 1985, pp. 50–52.

Wallich, Paul, "Toward simpler, faster computers," *IEEE Spectrum,* Vol. 22, No. 8, August 1985, pp. 38–45.

Wallich, Paul, and Zorpette, Glenn, "Minis and Mainframes," *IEEE Spectrum,* Vol. 23, No. 1, Jan. 86, pp. 36–39.

Wallis, P. J. L., "The Preparation of Guidelines for Portable Programming in High-Level Languages," *IEEE Computer,* Vol. 25, No. 8, August 1982, pp. 375–378.

Wang, Pearl S. C., and Kimbleton, Steve, "An Application Protocol for Networkwide Database Access," *Journal of Telecommunication Networks,* Vol. 2, No. 3, Fall 1983.

Wasserman, A., "The user software engineering methodology: An overview of interactive systems," *Information System Design Methodologies—A Comparative Review,* (A. Verrijn-Stuart, ed.) North Holland Pub. Co., Amsterdam, 1982.

Waters, Richard C., "The Programmer's Apprentice: Knowledge-Base Program Editing," *IEEE Transactions on Software Engineering,* Vol. SE-8, No. 1, January 1982, pp. 1–12.

Watson, Richard W., "Identifiers (Naming) In Distributed Systems," *Distributed Systems Architecture and Implementation* (B. W. Lampson, M. Paul, and H. J. Siegent, editors), Springer Verlag, 1981, pp. 191–210.

Weber, Herbert, "The Distributed Development System—A Monolithic Software Development Environment," *ACM SIGSOFT Software Engineering Notes,* Vol. 9, No. 5, October 1984, pp. 43–72.

Wegner, Peter, and Smolka, Scott A., "Processes, Tasks and Monitors: A Comparative Study of Concurrent Programming Primitives," *IEEE Transactions on Software Engineering,* Vol. SE-9, No. 4, July 1983, pp. 446–462.

Weicker, Reinhold P., "Dhrystone: A Synthetic System Programming Benchmark," *Communications of the ACM,* Vol. 27, No. 10, October 1984, pp. 1013–1030.

Weiderhold, Gio, *Database Design,* 2nd Edition, McGraw-Hill, New York, 1983.

Weiderhold, Gio, "Databases," *IEEE Computer,* Vol. 17, No. 10, October 1984, pp. 211–224.

Weitzman, C., *Distributed Micro/Minicomputer System: Structure, Implementation and Application,* Prentice-Hall, 1980.

Wenger, Peter, "Knowledge Engineering," *IEEE Software,* Vol. 1, No. 3, July 1984, pp. 33–38.

White, J. E., "A User Friendly Naming Convention for Use in Communication Networks," Proc. Working Conference on Computer Message Services, IFIP Working Group 65, May 1984, pp. 37–57.

Wienski, R., "Evolution to ISDN within the Bell Operating Companies," *IEEE Communications Magazine,* January 1984.

Williams R., Obermark, R., Daniels, D., Haas, L., Lapis, G., Lindsay, B., Ng, P., Selinger, P., Walker, A., Wilms, P., and Yost, R., "R*: An Overview of the Architecture," Proceedings of the International Conference on Database Systems, Jerusalem, Israel. Published in *Improving Database Usability and Responsiveness* (P. Scheuermann, editor), Academic Press, N.Y., June 1982, pp. 1–27.

Williams, R., and Gillman, R., "ISDN Access Protocols—Status and Applications," Proceedings, National Communication Forum 1984.

Williams, Tom, and Goering, Richard, "Data Communication Networks Finally Come of Age," *Computer Design,* Vol. 23, No. 14, December 1984, pp. 169–196.

Williamson, Ronald, and Horowitz, Ellis, "Concurrent Communication and Synchronization Mechanisms," *Software Practice and Experience,* Vol. 14, No. 2, February 1984, pp. 135–151.

Wilms, Paul F., and Lindsay, B. G., "A Database Authorization Mechanism Supporting Individual and Group Authorization," IBM Research Report RJ3137, May 1981. Also available as Distributed Data Sharing Systems (Van De Riet, R. P., and Litwin, W., editors), North-Holland Publishing Company, 1981.

Wilms, Paul F., Lindsay, B. G., and Selinger, P., " 'I Wish I Were Over There,' Distributed Execution Protocols for Data Definition in R*," Proc. SIGMOD 83, San Jose, CA, May 23–26, 1983, pp. 238–244. Also available as IBM Research Laboratory Report RJ3892, San Jose, California, May 1983.

Wilson, Pete, "Language-Based Architecture Eases System Design—I," *Computer Design,* Vol. 22, No. 13, November 1983, pp. 107–115.

Wilson, Peter, "Language-Based Architecture Eases System Design—II," *Computer Design,* Vol. 22, No. 14, December 1983, pp. 109–120.

Williamson, R., and Horowitz, E., "Concurrent Communication and Synchronization Mechanisms," *Software Practice and Experience,* Vol. 14, No. 2, February 1984, pp. 135–151.

Winski, Donald T., "Distributed Systems—Is Your Organization Ready?," *Infosystems,* Vol. 25, No. 9, September 1978, pp. 38–42.

Winston, Patrick H., *Artificial Intelligence* (2nd eds.) Addison-Wesley Publishing Company, 1984.

Winston, Patrick H., and Prendergast, Karen A., (eds.), *The AI Business,* The MIT Press, Cambridge Massachusetts, 1986.

Wirth, N., "Modula: A Language for Modular Multiprogramming," *Software Practice and Experience,* Vol. 7, No. 1, January 1977, pp. 3–35.

Wirth, N., *Programming in Modula-2,* Springer-Verlag, New York, 1983.

Witt, B., "Communicating Modules: A Software Design Model for Concurrent Distributed Systems," *IEEE Computer,* Vol. 18, No. 1, January 1985, pp. 67–77.

Wong, Eugene, and Katz, R., "Distributing a Database for Parallelism," Proc. of the ACM SIGMOD 83, International Conf. on Management of Data, San Jose, CA, May 23–26, 1983, pp. 23–28.

Woods, William R., "What's Important About Knowledge Representation?," *IEEE Computer,* Vol. 16, No. 10, October 1983, pp. 22–27.

Wu, Meng-Lih, and Hwang, Tai-Yang, "Access Control with Single-Key-Lock," *IEEE Transactions on Software Engineering,* Vol. SE-10, No. 2, March 1984, pp. 185–191.

Wulff, Robert S. M., "Multiple Micros Distribute Text and Graphics Functions," *Computer Design,* Vol. 23, No. 12, October 15, 1984, pp. 141–147.

Wurzburg, Henry, and Kelley, Steve, "PBX Based LANs: Lower Cost per Terminal Connection," *Computer Design,* Vol. 23, No. 2, February 1984, pp. 191–199.

Yalamanchili, S., Malek, M., and Aggarwal, J. K., "Workstations in a Local Area Network Environment," *IEEE Computer,* Vol. 17, No. 11, November 1984, pp. 74–86.

Yau, Stephen S., "Japanese Computer Technology and Culture," *IEEE Computer,* Vol. 17, No. 3, March 1984, pp. 4–5.

Yau, Stephen S., and Caglayan, Mehmet U., "Distributed Software System Design Representation Using Modified Petri Nets," *IEEE Transactions on Software Engineering,* Vol. SE-9, No. 6, November 1983, pp. 733–745.

Yost, Robert A., and Lindsay, B. G., "A Distributed Terminal Management Facility," IBM Research Report RJ3652, January 1983.

Yu, Clement T., Chang, C. C., Templeton, M., Brill, D., and Lund, E., "On the Design of a Query Processing Strategy in a Distributed Database Environment," in Proceedings of the ACM SIGMOD 83 International Conf. on Management of Data, San Jose, CA, May 23–26, 1983, pp. 30–39.

Yu, C. T., Guh, K., Chang, C., Chen, C., Templeton, M., and Brill, D., "An Algorithm to Process Queries in Fast Distributed Network," in Proceedings of the IEEE Real Time Systems Symposium, Austin, TX, 1984, pp. 115–122.

Yu, C. T., and Chang, C. C., "Distributed Query Processing," *ACM Computing Surveys,* Vol. 16, No. 4, December 1984, pp. 399–433.

Ziegler, K., "A Distributed Information Study," *IBM Systems Journal,* Vol. 18, No. 3, September 1979, pp. 374–401.

Zimmerman, H., "OSI reference mode—The ISO model of architecture for open systems interconnection," *IEEE Transactions on Communications,* Vol. COM-28, No. 4, April 1980.

Zorpette, Glenn, "Microprocessors," *IEEE Spectrum,* Vol. 22, No. 1, January 1985, pp. 53–55.

Zorpette, Glenn, "The Beauty of 32 bits," *IEEE Spectrum,* Vol. 22, No. 9, September 1985, pp. 65–71.

Zwaenepoel, Willy, and Lantz, Keith A., "Perseus: Retrospective on a Portable Operating System," *Software Practice and Experience,* Vol. 14, No. 1, January 1984, pp. 31–48.

Index

Index